Biomedical Perspectives of Herbal Honey

Rajesh Kumar • Suresh Kumar
Shamsher S. Kanwar

Biomedical Perspectives of Herbal Honey

Rajesh Kumar
Department of Biosciences
Himachal Pradesh University
Shimla, Himachal Pradesh, India

Suresh Kumar
Department of Biosciences
Himachal Pradesh University
Shimla, Himachal Pradesh, India

Shamsher S. Kanwar
Department of Biotechnology
Himachal Pradesh University
Shimla, Himachal Pradesh, India

ISBN 978-981-97-1528-2   ISBN 978-981-97-1529-9 (eBook)
https://doi.org/10.1007/978-981-97-1529-9

© The Editor(s) (if applicable) and The Author(s), under exclusive license to Springer Nature Singapore Pte Ltd. 2024

This work is subject to copyright. All rights are solely and exclusively licensed by the Publisher, whether the whole or part of the material is concerned, specifically the rights of translation, reprinting, reuse of illustrations, recitation, broadcasting, reproduction on microfilms or in any other physical way, and transmission or information storage and retrieval, electronic adaptation, computer software, or by similar or dissimilar methodology now known or hereafter developed.

The use of general descriptive names, registered names, trademarks, service marks, etc. in this publication does not imply, even in the absence of a specific statement, that such names are exempt from the relevant protective laws and regulations and therefore free for general use.

The publisher, the authors, and the editors are safe to assume that the advice and information in this book are believed to be true and accurate at the date of publication. Neither the publisher nor the authors or the editors give a warranty, expressed or implied, with respect to the material contained herein or for any errors or omissions that may have been made. The publisher remains neutral with regard to jurisdictional claims in published maps and institutional affiliations.

This Springer imprint is published by the registered company Springer Nature Singapore Pte Ltd.
The registered company address is: 152 Beach Road, #21-01/04 Gateway East, Singapore 189721, Singapore

Paper in this product is recyclable.

Foreword

Shiv Pratap Shukla
Governor
Himachal Pradesh

The book volume entitled *Biomedical Perspectives of Herbal Honey* compiled and edited by Dr. Rajesh Kumar, Dr. Suresh Kumar, and Prof. Shamsher S. Kanwar from Himachal Pradesh University, Shimla (India), to be published by Springer Nature is a compendium of different aspects of honey.

Medicinal herbs have extensively been used since time immemorial in traditional healthcare practices for the treatment of various diseases. The earliest record of the use of honey and medicinal plants for prevention as well as cure of diseases can be traced in Rigveda which is considered to be the oldest repository of human knowledge. In Ayurveda, the detailed properties and uses of honey and plant-based drugs have been mentioned. According to the reports of the World Health Organization, more than 80% of the world's population still relies on plant-based traditional medicines for primary healthcare management. In recent years, there has been growing interest in alternative therapies and the therapeutic use of natural products, particularly those derived from plants. There have been vigorous efforts globally to conserve, document, and promote the knowledge of plant-based drugs and to develop pharmacological research programs for the benefit of traditional and modern medicinal systems. Worldwide consumption of medicinal plants for various uses worth thousand crores rupees. It has been estimated that the market size of the medicinal plants-based industry is about 60 billion dollars annually.

Despite recent developments in synthetic drugs, the majority of people still depend upon traditional medicines because of their low cost, lowest side effects, and accessibility in remote areas. In the present scenario, rich herbal treasures and traditional knowledge are the key components for bio-prospecting, value addition, and research development. Generally, herbs are mixed with honey to make a paste or electuary that dates back to our ancient cultures where honey was used to preserve herbal formulations for longer periods. Presently, honey is added to different herbs like Tulsi, Ginger, etc. to cure several diseases. It is also used for culinary purposes and added to baked goods, salads, and marinades to enhance flavor as well as taste. Herb-infused honey is taken medicinally to cure a large number of diseases like cough, cold, fever, asthma, tuberculosis, jaundice, chest infections, urinary troubles, allergies, diabetes, liver disorders, sexual problems, cardiac disorders, wound healing, immunity enhancer, etc. It is the fact that honey is well known for its healing properties and its infusion with herbs further paves a tasty and healthy way for the treatment of various diseases.

I greatly appreciate the efforts put in by the editors for successfully bringing together this impressive volume and wishes them good luck.

Himachal Pradesh University, Raj Bhawan	Shiv Pratap Shukla
Shimla, Himachal Pradesh, India	

Preface

Use of herbs and honey as medicine has been in practice since human sustenance. These natural resources have been used for drugs as well as nutrition since prehistoric times. They remained a considerable part of lifestyle in all civilizations on earth. Still today, it has been estimated that about 80% of the global population cannot afford industrial synthetic products and hence rely on utilizing traditional medicines mainly derived from natural resources including plants and honey. The use of different parts of various medicinal plants has been in vogue from ancient times. In modern medicine also, plants occupy a very significant place as raw material for some important drugs. Plant-based drugs are being increasingly preferred in medical science. Honey is regarded as an excellent medicine all over the world in addition to its use as everyday food. It has antibacterial, anti-inflammatory, immune boosting, and wound healing properties, providing several health benefits.

There is a growing tendency throughout the world to shift from synthetic to natural products. Recent developments in drug discovery have attracted the attention of scientists and researchers throughout the world. Various natural resources are subjected to biological screening using biomedical technologies which have been leading to the discovery of different new sources of therapeutic agents possessing antimicrobial, antidiabetic, anticancer, hepatoprotective, anti-oxidative, anti-arthritic, anti-asthmatic, and anti-ulcer properties. The researches in the discovery of new drugs from herbs have resulted in the treatment of several ailments like cancer, diabetes, arthritis, respiratory disorders, nervous disorders, and gastrointestinal complaints.

This is unfortunate that the full potential of herbs as well as honey and their combination for biomedicinal purposes has not been scientifically explored. There is a great necessity to comprehensively explore the biomedical aspects of herbal honey which will prove to be a landmark in the betterment of human health. In view of the importance of herbal honey and the need of the day, this book will provide comprehensive and condensed information, covering many important and interesting aspects on the subject. It will serve as a useful companion for the students, researchers, teachers, practitioners, beekeepers, industrialists, collectors, cultivators, foresters, and nature lovers. It will be of immense use to those who are enthusiastically concerned with biomedicine, phytochemistry, and pharmaceutical industry.

Shimla, Himachal Pradesh, India	Rajesh Kumar
Shimla, Himachal Pradesh, India	Suresh Kumar
Shimla, Himachal Pradesh, India	Shamsher S. Kanwar

Contents

1	**Honey: Introduction, History, Composition, and Its Uses**	1
	1.1 Historical Background	2
	1.2 Types of Honey	4
	1.2.1 Wild Honey	4
	1.2.2 Apiary Honey	4
	1.3 Extraction, Processing, and Packaging of Honey	7
	1.4 Storage of Honey	8
	1.5 Composition of Honey	9
	1.6 Physical Properties of Natural Honey	10
	1.7 Purity Standards	11
	1.8 Uses and Benefits of Honey	12
	1.9 Health and Medicinal Benefits of Honey	13
	1.10 Value-Added Products	15
	1.11 Conclusion	17
2	**Pharmacological Properties of Honey**	19
	2.1 Introduction	19
	2.2 Pharmacological Properties	20
	2.2.1 Antibacterial Properties	21
	2.2.2 Antimicrobial Properties	22
	2.2.3 Immunomodulatory Properties	23
	2.2.4 Antioxidant Properties	24
	2.2.5 Inflammatory Properties	26
	2.2.6 Prebiotic Properties	27
	2.2.7 Antifungal Properties	27
	2.2.8 Antiviral Properties	28
	2.2.9 Anticarcinogenic Properties	29
	2.2.10 Immunity Boosting Activity	30
	2.2.11 Antidiabetic	31
	2.2.12 Antihypertensive	31

		2.2.13	Cardio-Protective Role of Honey.	32
		2.2.14	Wound-Healing Properties of Honey.	33

3 Herbs as Therapeutics and Healers. 35
- 3.1 Introduction 35
- 3.2 Traditional Use of Herbs 36
- 3.3 Herbs as Therapeutics 37
- 3.4 Antioxidant Properties. 38
- 3.5 Blood Purification 39
- 3.6 Healing Properties of Some Common Herbs. 41
  - 3.6.1 *Achyranthes aspera* L. 41
  - 3.6.2 *Asparagus adscendens* Roxb.. 41
  - 3.6.3 *Azadirachta indica* A. Juss. 41
  - 3.6.4 *Cannabis sativa* L. 41
  - 3.6.5 *Cinnamomum camphora* (L.) Nees & Eberm 42
  - 3.6.6 *Curcuma longa* L. 42
  - 3.6.7 *Glycyrrhiza glabra* L. 42
  - 3.6.8 *Ocimum sanctum* L. 42
  - 3.6.9 *Phyllanthus emblica* L. 43
  - 3.6.10 *Podophyllum hexandrum* Royle 43
  - 3.6.11 *Ricinus communis* L. 43
  - 3.6.12 *Taraxacum officinale* F.H. Wigg. 43
  - 3.6.13 *Thymus serpyllum* L. 44
  - 3.6.14 *Tinospora cordifolia* (Willd.) Hook.f. & Thomson 44
  - 3.6.15 *Withania somnifera* (L.) Dunal 44
  - 3.6.16 *Zingiber officinale* Roscoe. 44
- 3.7 Precautions for Herbal Treatment 45
- 3.8 Herbal Products 45

4 Herbs of Bee Interest 47
- 4.1 Introduction 47
- 4.2 Floral Calendar. 48
- 4.3 Bee Flora 49
- 4.4 Description of Some Common Honey Plant Resources 51
- 4.5 Honey Flow Sources 65

5 Herbs Honey Infusion Methods. 67
- 5.1 Methods of Making Herb-Infused Honey 67
  - 5.1.1 Infusion Method I (Grabek-Lejko et al. 2022). 67
  - 5.1.2 Infusion Method II (Putri et al. 2022) 68
  - 5.1.3 Infusion Method III (Ewnetu et al. 2014) 69
  - 5.1.4 Infusion Method IV (Tomczyk et al. 2020) 71
  - 5.1.5 Infusion Method V (Miłek et al. 2023) 72
  - 5.1.6 Infusion Method VI (Ewnetu et al. 2014) 72
  - 5.1.7 Infusion Method VII (Jafari et al. 2023) 73
  - 5.1.8 Infusion Method VIII (Laksemi et al. 2023) 73

6	**Herbal-Infused Honey vis-à-vis Human Health**		**75**
	6.1	Introduction	75
		6.1.1 Herbal Therapy for Disease Treatment	76
		6.1.2 Herbal Medicine Use in COVID-19 Management	76
	6.2	Herbal Infusions: Harmony in Therapeutic Potential	78
		6.2.1 Herbs Infused with Honey	78
		6.2.2 Health Benefits of Honey	79
		6.2.3 Honey in Herbal Medicine Formulations	81
	6.3	Study of Synergism of Honey with Various Herbs	81
		6.3.1 Allium sativum L	81
		6.3.2 Alpinia officinarum	83
		6.3.3 *Capparis spinosa* L	84
		6.3.4 *Curcuma xanthorrhiza* Roxb	84
		6.3.5 *Euphorbia hirta* L	85
		6.3.6 *Nigella sativa* L	86
		6.3.7 *Phyllanthus emblica* L	87
	6.4	Conclusion	88
7	**Biomedical Perspectives of Herbal Honey**		**89**
	7.1	Herbal Honey in Communicable Diseases	90
		7.1.1 Honey and COVID	90
		7.1.2 Herbal Honey and Dermatitis	93
		7.1.3 Herbal Honey and Eye Disorders	101
		7.1.4 Herbal Honey and Hypersensitivity	107
		7.1.5 Herbal Honey and Influenza	112
		7.1.6 Herbal Honey and Periodontal Disease	117
		7.1.7 Herbal Honey and Herpes Disease	118
		7.1.8 Herbal Honey and Viral Hepatitis	120
		7.1.9 Herbal Honey and Gingivostomatitis	122
		7.1.10 Herbal Honey and Anal disorders	125
	7.2	Herbal Honey in Non-Communicable Diseases	127
		7.2.1 Herbal Honey and Diabetes	127
		7.2.2 Herbal Honey and Cancer	130
		7.2.3 Herbal Honey and Wound Management	142
		7.2.4 Herbal Honey and Alzheimer's Disease	148
		7.2.5 Herbal Honey and Gastrointestinal Disorders	152
		7.2.6 Treatment of the Gastrointestinal Disorder	154
		7.2.7 Herbal Honey and Cardiovascular Disorders	156
		7.2.8 Herbal Honey and Reproductive Disorders	162
8	**Critical Analysis and Future Perspectives**		**169**
	8.1	Challenges	171
	8.2	Future Perspectives	172
	8.3	Recommendations	172
Bibliography			**173**

About the Editors

Rajesh Kumar Dr. Kumar did his Masters in Zoology from Jiwaji University, Gwalior (M.P.), India, in 2007 and awarded Ph.D. from the same institution in the year 2013. Dr. Kumar started his academic career in 2013 and currently working as an Assistant Professor in the Department of Biosciences, Himachal Pradesh University, Shimla (Himachal Pradesh), India. He has about 10 years of experience in teaching and research. Dr. Kumar pursues his research in Animal Physiology and Applied Zoology. Dr. Kumar has completed 02 major research projects and 02 startup projects sponsored by Government agencies. He has successfully supervised 03 Ph.D., 01 M.Phil., and 43 postgraduate students for their research work. Dr. Kumar has been awarded a Meritorious Fellowship for the year 2009–2011 by the University Grants Commission, Govt. of India. He is a Fellow of Himalayan Science and Technology Communications and is a life member of many prestigious agencies. He has published more than 125 research articles including 50 book chapters in indexed journals and publishers such as Elsevier/Springer, 12 articles in full-length proceedings, 16 books (Springer Nature, CRC Press, AAP), 04 monographs, about 115 abstracts by participating in more than 80 national/international conferences. He has earned 12 awards from reputed organizations for his excellent research. Dr. Kumar has successfully organized many seminars/conferences and is actively involved in delivering invited talks/extension lectures at various events. He is also a reviewer of several journals of national and international repute. He has Google Scholar 700+ citations, H-index of 11, and i10 index of 13.

Suresh Kumar is presently working as an Associate Professor in the Department of Biosciences, H.P. University, Shimla. He completed his doctoral degree from H.P. University, Shimla, in 2010. He has more than 15 years of experience in teaching as well as research. His thrust areas include Floristics, Ethnobotany, and Biodiversity. He qualified UGC-CSIR (NET) in 2002. During Ph.D., he was awarded a Junior Research Fellowship for the period 2005–07 and a Senior Research Fellowship for the period 2007–09 by UGC. Dr. Kumar has published 15 books and more than 20 research papers in various journals of National and International repute. Presently, he has been guiding 07 students for Ph.D. degrees and 04 students for M.Sc. projects. He has participated in many Conferences/Seminars and delivered invited talks as a resource person in various institutions. He is a member of the Indian Science Congress and many other scientific agencies. He is an Associate Editor as well as a Reviewer of various journals in the field of Botany. He has more than 260+ Google citations with h-index=8 and i10 index=6.

Shamsher S. Kanwar Dr. Kanwar is Senior Professor at the Department of Biotechnology, Himachal Pradesh University, Shimla (India), does research in nanotechnology, medicinal chemistry, therapeutics, and biocatalysis. One of his current projects is "Improvement of activities of industrially important enzyme by protein engineering, rational design and enzyme immobilization." Dr. Kanwar has guided ~175 students for their postgraduate degrees in Biotechnology with dissertation(s), 30 candidates for M.Phil. in Biotechnology, and 21 candidates for Ph.D. in Biotechnology. He pursues his research activities by developing bio-products and bio-processes for a variety of extracellular microbial enzymes to assess the antitumor and anticancer activities of purified microbial enzymes. Dr. Kanwar has published 240 articles including chapters in the books (published by Elsevier, Willey, Springer, InTech, CRC Press, Taylor & Francis, etc.) over a career spanned over 30 years. He has filed 06 patents, all these have been published, and one of these has been granted. Dr. Kanwar is a certified reviewer of 56 international and national peer-reviewed journals. He is also editor of Current Biotechnology, Trends in Oncology, Journal of Advanced Microbiology, Insight in Enzyme Research and Current Research in Virology and Retrovirology. He has Research Gate Score of 77.09, Google Scholar 6400+ citations, H-index of 35, and i10 index of 102.

Chapter 1
Honey: Introduction, History, Composition, and Its Uses

Honey is one of the miraculous products of beekeeping ventures prepared by honeybees. It is considered one of the wonder products out of all-natural harvests available in the universe for the welfare of mankind. Honey is one of the most sweet, delicious, and nutritious substances, viscous in appearance and having a wide range of colors. Honeybees collect nectar from nectarines of flowers from different types of crops, viz. crop, medicinal, orchard, vegetable, fruits, and ornamental as well. They fill nectar by regurgitating it after adding various enzymes. It is called mature nectar or raw honey and usually less sweet and watery in appearance. Honeybees transform it into honey by fanning their wings, making it viscous, and finally sealing it in the comb cells with the help of wax. Honey is mainly a saturated solution of sugars along with a rich composition of various enzymes, minerals, vitamins, and other nutritious elements. It is one of the most important beekeeping products from a nutritive and commercial point of view. It is a wonderful appetizer, assimilator, energizing, and tasty product that serves as an excellent diet, tonic, supplement, medicine, and cosmetic agent without causing any harm to the body.

Honey is the first-ever sweet product used as a source of energy by human beings since time immemorial. The evidence for the use of honey can be seen among all human civilizations and in all spheres of life, i.e., medicinal, religious, and rituals. It is used at the birth of young ones as well as in death ceremonies in many countries such as India. The description of honey can be found in all religious books (Gita, Bible, Quran, and Vedas). In Indian Vedas, which are considered the oldest religious scriptures and storehouse of knowledge, honey has been given special mentions.

The importance and potential of honey have been well recognized by our forefathers. Even today, honey is given to neonatal as the first food, prior to mother's milk. In Sanatan Dharma, honey is added as one of the ingredients to prepare "Panchamrut" (a mixture of milk, curd, ghee, holy water from Ganga, and honey). Christians use honey in certain religious ceremonies. The Quran has a special mention of honey and its uses. Honey has been used as a sweetening agent in worshipping, rituals, and preparation of dishes.

In the Ayurvedic system of medicine, honey has been recognized as a multipurpose remedy for various diseases and a carrier/assimilator (Yogavahi) of medicines. The use of honey as a food source is an age-old practice, an extraordinary place in the nutraceutical industry due to the presence of natural sugars and the only source of sweetener used by the common man since ancient times. With the advancements in technology and scientific interventions, honey has been used to prepare many value-added products that are of more importance and have higher nutrition as well as commercial importance.

Honey produced from various honeybee species is used in traditional folk medicine systems, i.e., Ayurveda, Unani, Siddha, and Homeopathy, as an assimilator, absorbent, and preservative. Due to magnificent properties like sweet taste, palatability, digestibility, blood purification, antioxidant, antibacterial, antitussive, and antifungal, honey is used as a promising food supplement and part of medicine all over the world. Honey aids in digestion, increases appetite, and helps in the assimilation of food/energy inside our body that ultimately regulates metabolism. It is also helpful in the management/treatment/control of various chronic diseases like renal, hepatic, pulmonary, etc. and is a proven remedy against cough, cold, fever, and other microbial infections (bacterial, viral, and fungal). Honey possesses wonderful rejuvenating properties, due to which it is used to nourish skin in the form of face packs, creams, etc. Undoubtedly, due to all these qualities, honey is a unique gift of nature.

1.1 Historical Background

The history of honey and beekeeping is as old as human civilization. It dates back to the Cretaceous period when people lived in caves, flowering plants evolved on the Earth, and honeybees also evolved in nature at the same time. Honey was supposed to be the food of wild animals first, however, with time, primitive man (Adimanav) also came to know about the edible properties of honey and started honey hunting. Explanation of honey is presented even in ancient literature of all cultures around the globe. It is mentioned as the food of gods. Honey pots were reported in Egyptian mummies, among other materials and utensils. The Indian Epic Rigveda (2000–3000 BC), one of the oldest scriptures has mentions of honey and honeybees. The earliest bees appear to have originated in the xeric interior of the supercontinent Gondwana, the region for the origin of flowering plants. This indicates the presence of bee flora for more than 90 million years. Adequate fossil records are not available to trace the exact ancestral phylogenetic line. However, a description of a specimen of the sphenoid wasp from the Cretaceous period links it to a possible ancestor of bees. Thus wasp-like ancestors of honeybees took advantage of the food provided by flowers and began to modify their diet and physical characteristics. Literature suggests that bees have largely acted as cross-pollinators since geological times.

1.1 Historical Background

The first mention of honeybees and honey dates back to around 2400 BC, in an official list of beekeepers. Aristotle around 384-322 BC observed the persistence of bees on the flowers of specific crops and also mentioned flower fidelity, division of labor, and diseases in honeybees. The earliest evidence of beekeeping by hives have reported from the old kingdom of Egypt.

Although the economic importance of honeybees has been known to man since the pre-civilization era, scientific beekeeping began over 500 years ago. Beekeeping originated and developed between 1500 and 1851 when efforts were made to rear honeybees in different types of hives for the purpose of commercial purposes. A significant contribution was made by F. Huber from Geneva who conducted the first scientific study of bees during 1792.

Modern beekeeping started in the late nineteenth century when L. L. Langstroth designed and developed wooden boxes with movable frames for rearing honeybees (Fig. 1.1). This was a major breakthrough in the field of beekeeping that led to the establishment of commercial beekeeping later. Subsequently, J. Mehring designed artificial comb foundation sheets of wax for their use in commercial beekeeping.

The commercialization of honey was popularized and flourished during the twentieth century when the properties of honey were explored and the common man came to know about the importance of honey. Nowadays, beekeeping has grown as the most popular rural cottage industry in many countries. People are adopting beekeeping as their hobby as well as a full-fledged source of income. Beekeepers are now entering the area of other valuable products of honeybees apart from honey as well as industries are coming up for manufacturing value-added products from beekeeping products.

Fig. 1.1 Langstroth beehive

1.2 Types of Honey

Honey has been categorized into various types depending on the kind of floral sources, from which honeybees extract nectar. Although the basic composition of honey remains the same, however, taste, aroma, sugar content, moisture content, and some of the medicinal properties may differ as per the dominance of flowers available for honeybees. There are mainly two types of honey:

1.2.1 Wild Honey

Honeybee species that inhabit natural conditions i.e. forests, rocks, open areas, etc. are capable of producing ample quantities of honey which is termed wild honey. Among all of the wild bee species, Apis dorsata is the largest producer of honey however, it is not possible to harvest honey from all bee hives that exist in wild. Wild honey is equally tasty, nutritious, and medicinal as that of apiary honey. It has high demand in local markets all over the world. Sometimes it is said that wild honey has more therapeutic value that apiary honey which has no scientific evidence.

1.2.2 Apiary Honey

1.2.2.1 Wild Honey

Honey obtained from wild honeybee species, *Apis dorsata*, and dwarf bee, *Apis florea* is generally called wild honey. These honeybee species predominantly occur in forest areas and are the largest producers of honey. However, complete honey cannot be extracted from *A. dorsata* combs due to various factors such as comb distribution in forests and traditional honey hunting methods.

Apis dorsata make their hive on trees, rocks, or buildings. Honey is extracted using the traditional honey hunting method, bees are chased away by the smoke by setting fire, and the portion of the comb is cut down killing a large number of bees and brood. The cut-out portion of comb-carrying honey is kept in a cloth and squeezed to bring out honey. In this kind of unrefined "squeezed honey," bee eggs, brood, pollen grain/bee bread, wax particles, parts of dead bees, dust, etc., are largely present. Also, it has a high moisture content; hence, it cannot be preserved for a long period and usually turns black in a short period. Due to all these factors, wild honey has a low market value and is not sold commercially by any brand company. *Apis florea* on the other side prepares its combs in bushes at lower places. It produces a non-significant quantity of honey that cannot be extracted for commercial purposes (Fig. 1.2).

1.2 Types of Honey

Fig. 1.2 (i) Wild honeybee comb, (ii) Apiary

1.2.2.2 Apiary Honey

Honey obtained from *Apis cerana* and *Apis mellifera* reared in wooden bee boxes and extracted through scientific methods is called apiary honey. Both these species are of gentle temperament and produce good-quality honey which is pure and clean. Worker bees are used to fill honey inside comb cells and seal it with the help of wax. During extraction, honey-filled frames, devoid of eggs and brood, are brought out of the hive and rotated by placing them in a honey extractor. Honey comes out of the comb cells due to centrifugal force, and there is no damage to the comb/frame. In this way, "extracted honey" has the least amount of water and pollen, bee bread, wax particles, etc. It can be further processed and preserved for a longer period of time under favorable conditions. This type of honey is made available in the market in packaged form under different brand names.

Apiary honey may be categorized into different types as follows:

(i) Multifloral honey
(ii) Unifloral honey
(iii) Cream honey
(iv) Comb honey

(i) **Multifloral Honey:** Honeybees prepare honey by collecting nectar from the flowers. They are free-wandering insects and forage upon all kinds of flowers of their interest. While visiting flowers, honeybees collect their nectar and transform it into honey. Hence, a variety of nectar is stored and this aids in the production of multifloral honey having taste, aroma, and properties of multiple flowers from which the nectar has been collected. Most of the honey available in the market is of a multifloral nature.

(ii) **Unifloral Honey:** This type of honey can be obtained when an apiary is established in a particular type of flower-dominated area, i.e., the same type of flow-

Fig. 1.3 Apiary in apple orchard

ers are in abundance. The majority of honeybees shall bring in nectar from the same type of flowers, hence unifloral types of honey can be obtained from the hives, e.g., neem honey, litchi honey, mustard honey, sunflower honey, citrus honey, eucalyptus honey, apple honey, almond honey, and cashew honey, etc. (Fig. 1.3).

(iii) **Cream Honey:** Pure honey sometimes gets granulated especially in the areas where the temperature is low. Such a type of honey is called cream honey or granulated honey. It is eaten like butter or cream by spreading it on bread, toast, slices, biscuits, cakes, etc. This type of honey is in great demand in many countries.

(iv) **Comb Honey:** Honeycomb pieces filled with honey are called comb honey. Many people prefer to extract and use honey from honeycomb pieces themselves, thereby avoiding the chances of adulteration. Some people directly suck and eat the pieces of honeycomb filled with honey and throw wax out of their mouth like the kernels of a fruit. The demand for this type of honey is increasing in the market these days (Fig. 1.4).

(v) **Seasonal Honey:** In a few countries like India, honey has been categorized on the basis of the season as Basant Honey (extracted in the Spring season), Summer Honey (extracted in summer), Kartik Honey (extracted in the autumn season), and Sharad Honey (extracted in winter season).

(vi) **Organic Honey:** Honey free from all types of chemicals is called organic honey. However, truly organic honey is very rare to find. Different types of chemicals or their traces get incorporated in honey through pollen, nectar, or even with body parts of honeybees. Honey can be considered organic only if the amount of chemicals is below the standard value. Sometimes the unscien-

1.3 Extraction, Processing, and Packaging of Honey

Fig. 1.4 Comb honey

tific method of extraction or poor storage conditions (tin, iron containers, etc.) may also lead to adulteration in honey. Beekeepers in many developing countries are either illiterate or have the least knowledge about these aspects; hence, traces of chemicals, heavy metals, pesticides, or their traces enter honey. Such honey is called adulterated or contaminated honey and has a poor market at the national or international level as it fails in purity checks. A few years ago, Indian honey was defamed in European countries after it failed purity checks and was declared sub-standard.

1.3 Extraction, Processing, and Packaging of Honey

Cleanly extracted, mature, and fresh honey is best and does not require any refinement. However, processing and packing of honey are necessary to remove foreign substances from honey, to preserve the quality of honey for a long time, and to prevent fermentation and granulation, so as to increase the market value. Manually, honey can be freed from particles of wax, pollen, dust, and other unwanted particles by passing through net clothes after heating at a controlled temperature (40 degrees Celsius for 2 hours) in the water bath. To pasteurize honey, a process of rapid heating and cooling is adopted. In many developing countries, beekeepers cannot afford the cost of expensive processing equipment. They heat the honey in whatever available containers with them and pack it after filtering it through cloth after cooling. Such unscientific processing, heating, and packing adversely affect the color, aroma, taste, and quality of honey and decrease its market value. With the advancements in technology, the quality and processing potential of honey extractors, processing, and packaging have also been improved. In advanced types of honey processing

Fig. 1.5 Honey stored in air-tight glass containers

plants, there is also a provision to reduce or maintain moisture level which is one of the main parameters to be controlled before packing the honey in containers. Air-tight glass containers should be used to store honey for a longer time (Fig. 1.5).

1.4 Storage of Honey

Honey gets darkened in color and granulates when it is stored for a long period of time. Fermentation also takes place due to chemical reactions between broken-down unstable levulose, amino acids, and tannates with iron salts. The minor color change is also due to the action of acidic honey on the metallic covers of storage containers. The sugars present in honey crystallize at low temperatures, initially give hazy appearance to liquid honey, and eventually solidify into granules. Granulation takes place at a faster rate if air bubbles, pollen, and dust particles are present in larger amounts. Therefore, honey should be heated at about 70 degrees Celsius for 30 minutes so that air bubbles disappear and crystals, if any may dissolve. Heated honey usually does not get crystalized; however, under severely colder conditions, granules can be seen at the bottom and liquid honey keeps on floating on the surface. Certain sugar-tolerant yeasts are present in the air which gets mixed with honey because of one or another means that contaminates honey. In fully mature/ripened honey, these yeasts are not able to grow because of the presence of high sugar concentration. Unripe honey having more than 20% moisture is the most favorable medium for the growth of such yeasts. These yeasts are responsible for the generation of alcohol by acting on sugars present in honey. Eventually,

1.5 Composition of Honey

Fig. 1.6 Mature honey frame and freshly extracted honey

a fermentation process takes place that causes the release of acetic acid, water, and high acidity. Such fermented honey turns blackish and sour. To prevent fermentation, only ripened honey should be extracted from the frames with the least moisture content and heated to 70 degrees Celsius for half an hour, and should be bottled in air-tight glass containers (Fig. 1.6).

1.5 Composition of Honey

Honey, the most precious edible item is obtained from honeybees by man from their nests. It is regarded as *Ambrosia,* the food of gods. Nutritionally, one tablespoon of honey (about 21 grams) contains 17 grams of sugars like fructose, glucose, maltose, and sucrose, and it has 64 calories of energy. Chemically, it has approximately 38.2% fructose, 31.3% glucose, 7.1% maltose, 1.3% sucrose, 17.2% water, 1.5%

higher sugars like maltodextrin, 0.2% ash, and up to 3.2% other nutrients. Other nutritional contents of honey are vitamins like ascorbic acid, pantothenic acid, niacin, and riboflavin; along with minerals such as calcium, copper, iron, magnesium, manganese, phosphorus, potassium, and zinc, antibiotic-rich inhibin, many amino acids, proteins, phenol antioxidants, and other micronutrients. Honey also comprises 4 to 5% fructo-oligosaccharides, which makes honey a strong probiotic agent. Water is the second most important component of honey. Organic acids present in honey contribute to its acidity and taste. Honey also has a large number of enzymes such as invertase, diastase, oxidase, glucose oxidase, invertase, amylase, catalase, etc. One of the substances, hydrogen peroxide which makes honey antimicrobial is formed from the enzyme glucose oxidase, combines with gluconic acid produced from glucose, and contributes to the absorption of calcium.

Fresh honey also has impurities such as wax, remains/parts of bees, brood, pollen, etc. Sometimes, heavy metals and pesticides also get fortified in honey. Some impurities appear in honey during extraction, filtration, and packaging. These impurities are not removed, or honey is not processed properly as per standard protocols.

1.6 Physical Properties of Natural Honey

Honey possesses diverse properties that make it distinct from all other products. The color of honey is the first and most attractive physical property that attracts consumers. Generally, light-colored honey is preferred in the market and has higher acceptance in comparison to dark-colored honey. The color depends upon the type of

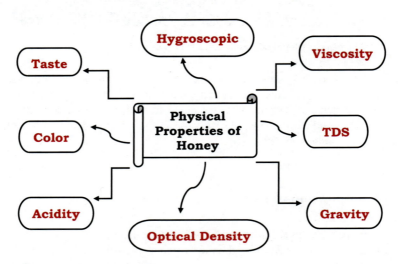

Fig. 1.7 Physical properties of honey

1.8 Uses and Benefits of Honey

flora, climate, and storage conditions. Viscosity is another important property of honey that depends upon various substances, especially water content. Fresh honey is generally a viscous fluid. The hygroscopic nature of honey causes it to adhere to moisture, normally it is below 18%. An increase in moisture changes the color of honey to darker and also deteriorates its taste. Apart from these, density, gravity, surface tension, total suspended solid, optical density, and many such properties are present in honey (Fig. 1.7).

1.7 Purity Standards

Honey cannot be tested for impurities or adulteration by consumers on their own. It can be checked in the laboratory only using various quantitative protocols. The best thing to do is to procure honey from a reliable source or reputable beekeeper. The popular notions that pure honey is not eaten by dogs or does not burn readily have no scientific basis. Homogenous granulation of honey is another sign of purity, which otherwise general masses consider as impure. Liquid honey free from dust, cloudiness, propolis, and parts of bees taken from reliable sources is also termed pure.

Lots of laboratories are coming up to check the purity of honey with the technological advancements. The only solution is to make the beekeepers aware of the relation between purity and commercial value, so that they must perform all

Fig. 1.8 Benefits of honey

activities related to honey venture using scientific interventions. This may help to raise the value as well as the standard of honey in the international market.

1.8 Uses and Benefits of Honey

Honey is very useful as a nutritious diet, health-promoting tonic, and medicine. It is mainly used in traditional folk medicinal systems and as a home remedy for treating various diseases (Fig. 1.8). The number of people using honey as food is negligible in developing countries but it is used as a food item (table honey) in many developed countries.

Honey is a natural, sugary, delicious, instantly refreshing complete food. The taste of honey is very sweet and palatable due to the sugars present in it, mainly levulose (fructose). The sugars in honey are in their simple form, glucose, and fructose. These sugars do not require any digestion process when taken in the form of food and they provide physical energy by being absorbed in the blood through the alimentary canal. Honey is a mini powerhouse packed with energy (calories) in the form of sugars. There are about 3500 calories in one kg of honey or about 100 calories in one teaspoon of honey. Honey can be consumed as desired, by applying it on bread, biscuits, etc. in water, milk, tea, coffee, sherbet, fruit juice, salad, fruit chaat, custard, rice, kheer, etc. in other foods like this, or with sugar.

Honey removes physical and mental fatigue and gives instant energy to the body. It is especially beneficial for newborn babies, growing children, old people, weak, disabled, sick people, pregnant women, sportsmen, athletes, climbers, divers, and people doing hard physical labor.

Honey can be consumed by all age-groups, i.e., from newborns to old and weak people. Some people may be allergic to honey (if not processed properly for removal of pollen), such people should not consume honey without medical advice. There is no fixed amount of honey intake, 1–2 teaspoonful of honey can be consumed daily or a dietician may be consulted to fix daily intake. Like the quantity of honey, there is no set time for consuming honey. It can be taken anytime in the morning on an empty stomach, before, after, or with food. Similarly, there is no certain way of consuming honey. It can be taken directly without dilution, mixed with water/milk (cold/lukewarm), juices, or in any other form. In many countries, frozen cream honey is eaten by spreading it on bread slices, toast, biscuits, etc. The smell and taste of honey in cold, uncooked foods do not change, and the smell and taste of some foods are enhanced by the addition of honey, such as lemonade, orange, other fruit juices, syrups, squash, jams, jellies, chutney, spread, ketchup, etc. Honey can be used in place of sugar in tomato sauce or ketchup.

Instead of adding honey to hot ketchup, it is better to add it when it is cold so that its properties may not get altered. Honey is twice as sweet as white sugar, so honey can be used in place of sugar, with or without sugar, in all kinds of foods. Honey can also be used with butter, cream, curd, buttermilk, cream, kheer, custard, jam, jelly, etc.

Honey is one of the oldest natural sweeteners and medications used since ancient times to treat infectious as well as microbial diseases. Recent studies have found that honey is a general immunity booster to fight against many diseases. Due to growing curiosity in the antimicrobial efficacy of honey, lots of researchers are getting attracted to it. Recent findings suggest that honey has therapeutic properties against micro-organisms, i.e., viruses, bacteria, fungi, etc.

Even though bees produce honey all over the world, its properties depend mainly upon botanical sources, other factors like processing, storage, and environmental conditions may also affect the quality/properties of honey to some extent. Due to the enormous variation in the compounds of honey, the therapeutic properties of honey are significantly multifaceted as a result of the mechanisms of many compounds. The major biological properties recognized as therapeutic are "antimicrobial" (bactericidal and fungicidal), "antiviral," and "antioxidant" activity. Besides this, honey is a naturally sweet, and non-toxic product that can be used as an immunity booster, wound healer, skin disinfectant, and to treat medical conditions like diarrhea, asthma, tumors, ulcers, and diabetes.

The benefits of honey have been recorded in ancient scriptures since ancient times. The composition of honey includes glucose, fructose, and minerals, i.e., Mg, Ca, K, NaCl, Fe, and P. Honey contains vitamins (B1, B2, B3, B5, B6, C), depending on the quality of nectar and pollen.

1.9 Health and Medicinal Benefits of Honey

Honey is considered a miraculous natural product that gives better physical and mental health and removes fatigue. Both healthy and sick people can use honey in their routine or for any disease. Honey can improve the growth of a newborn who is not breastfed and improve the calcium fixation of bones. Some of the significant health benefits of honey are as follows:

(i) **Honey for Digestive System**

Honey improves the absorption of food and benefits chronic and infective intestinal problems, including duodenal ulcers, constipation, and liver dysfunction. It has been reported that honey can treat gastrointestinal disorders.

(ii) **Honey and Respiratory System**

Honey is proven as a universal remedy against upper respiratory tract infections caused by polluted environments, climate change, and temperature fluctuations. Honey is believed to soothe and relax throat irritation and associated infections.

(iii) **Honey and Skin/Wound Healing**

Honey is used in different formulations that can be used straight away to open wounds, sores, multiple ulcers, and blisters. It is beneficial in skin rejuvenation and tissue repair/growth, and also reduces scars on the skin. It reduces the burning sensation and helps in the rapid formation of new tissues.

(iv) **Honey and Eyes**

Honey is effective in reducing and treating many ailments of the eyes, cataracts and swelling, and redness of the eyes. It is a proven remedy for eye flu (conjunctivitis) by controlling the bacterium *Pseudomonas aeruginosa*. In countries like India, honey is applied to the eyes with a needle during nighttime which helps in cleaning the eyes and keeping vision fine. Initially, it may cause some burning sensation, however, mixing rose water with honey reduces this problem and soothes the eyes. Mixing various herbs like clove and cardamom with honey may increase its benefits many folds.

(v) **Honey and Diabetes**

There are many claims that honey is beneficial for people suffering from diabetes. Honey is considered superior to the products synthesized from cane sugar/sugar. It has been found that insulin levels are lower in comparison to consuming other sweet foods of similar caloric values in honey. In a healthy person, regular consumption of pure honey maintains blood glucose levels than consuming the same amount of sucrose.

(vi) **Honey and Ayurveda**

In the Ayurvedic medicine system, honey is mainly used for the assimilation of various medicines such as herbal extracts. Apart from this, it is also known to be operative in treating various diseases. It may be used by people of all age-groups, convalescents, and as general restorative by hard-working laborers.

(vii) **Honey and Blood**

Consuming honey is beneficial in removing the deficiency of blood (hemoglobin) caused by weakness and disease. Taking honey with a glass of milk before bed may increase hemoglobin in the blood. Mixing various herbs like fennel, tulsi, ginger, etc. may help in the purification of blood.

(viii) **Honey in Physical Weakness**

Consuming honey by mixing it in water during the morning proves helpful in ending physical weakness and also gives relief in heaviness and pain in the body. It also removes mental tiredness, makes the mind cheerful, and also aids in memory.

(ix) **Honey and Insomnia**

Honey may prove beneficial in insomnia especially when taken at bedtime, it calms the nerves and brings sound sleep. Mixing lemon and honey in lukewarm water also leads to healthy and sound sleep.

(x) **Honey and Sex Problems**

Honey is an easy, simple, and best remedy for problems related to women's health, beauty, and sex. Continuous intake of honey repairs weakness, irregular menstrual cycles, menstrual pain, post-delivery health problems, lactation, leucorrhea, etc.

1.10 Value-Added Products

A large number of value-added products have been formulated to strengthen the honey industry. These products are tasty and nutritionally balanced which may prove helpful in maintaining good health. A few of them have been described as follows:

(i) **Honey Wine/Beer**

Mead is a honey wine, and its quality and flavor depend on the fermentation control and the properties of several ingredients, mainly on the characteristics and flavor of the honey. Good quality honey should be selected and for quality preparation, a good water supply should be maintained. Water can control the flavor of mead, mainly because water contains many minerals, chemicals, and other useful substances. Boiled water should be used for infection-free preparation.

The amount/ratio of ingredients to make mead mainly depends on the essential sugar water content of the honey and its alcohol content. Typically, in the final product, 2.3 kg of honey/100 L of water is assumed for each alcohol grade. More precisely, 21% solid sugar should be added to obtain a dry mead with 12% alcohol. When the solid sugar content is increased to about 25%, this leads to a final alcohol content of 14–15% and produces a sweeter mead. Pasteurization is an important step in the preparation of mead. Generally, it is not mandatory before fermentation. However, filtration is recommended to remove any solid granules. Various salts and minerals, including urea, ammonium, phosphate, cream of tartar, tartaric and citric acid, may be added as yeast nutrients. The acid improves the flavor of the mead and inhibits the growth of unwanted microorganisms. The production of mead is a month-long process, so that the mixture undergoes proper fermentation. The final step of mead preparation requires the addition of precipitating mediators including tannins, bentonite colloidal protein solution, or egg whites. Then the liquid mixture is filtered, and the final product is obtained. Compared to mead, the process of making honey beer is simple. However, honey beer cannot be stored for long periods of time and is suggested to be consumed fresh.

(ii) **Honey-Dry Fruits**

A variety of dried fruits that have little moisture and softness can be used with honey. They can be used in any form such as whole, sliced, and powder. Excess moisture from fruits should be removed otherwise it may initiate fermentation or increase hydroxy methyl furfural (HMF) content of honey turning the preparation sour.

(iii) **Honey Spread**

Honey spread can be prepared simply by keeping honey-filled containers in refrigerators. It can be eaten by spreading it on bread, toast, biscuits, cookies, cakes, and other products.

(iv) **Honey Infused Tea**

It is a famous type of tea already in use in many countries. Honey is mixed with other herbs like lemon, tulsi, and ginger to produce tea of flavors such as lemon honey, tulsi honey, and ginger honey. This type of tea is sold by many national and international brands and is easily available in the market.

Honey lemon tea can also be prepared at home as follows: Boil four cups of water in one spoon of tea and filter it into cups. Add the juice of fresh lemon

Table 1.1 Uses and products of honey

Product	Description
Wild turmeric soap-free antiseptic face wash	Restores natural balance without stripping skin of its natural moisture suitable for all skin types, the goodness of wild turmeric and honey gently cleanses by lifting all impurities and pollutants
Melia-face wash for oily skin and acne	It is a mild formula developed especially for oily skin, active ingredients like honey aloe vera and margosa dissolve oil and protect skin against germs and bacteria. It removes dirt from the deepest pores evens skin tone and refreshes skin
Punarnava detoxifying toner and astringent	Enlivens skin by deeply cleansing it and removing excess oil. A blend of herbal extracts like honey and almond oil, it helps in minimizing enlarged pores and maintaining elasticity giving the face a clean, clear, and radiant look
Protein-nourishing moisturizing cream	It is a blend of herbal extracts like milk, protein, almond oil, and honey. It restores the moisture level of the skin and also assists in the cell renewal process making it healthier, smoother, and visibly more radiant
Lotion- nourishing moisturizing lotion	It acts as a softening agent that improves skin texture and complexion. A light but rich emulsion containing fruit extract and honey that leaves the skin feeling resilient and refreshed
Surya sandal 20-sun block lotion	It is a mind moisturizing sun block with active ingredients like sandalwood and honey. It gives sun protection while hydrating the skin and giving it a refreshing after-feel
Slimmer's honey	It contains micronutrients and diet fortifiers, Triphala, detoxifies and tones your internal organs while pipli helps in digestion and respiratory decongestion. Other ingredients like ashwagandha promote vitality and strength while curing blood pressure
Morning nectar nourishing lotion	A light emulsion enriched with the natural goodness of honey, wheat germ, and seaweed. This lotion is easily absorbed, replacing the natural oils and moistures the skin loses every day
Dabur Honitus cough syrup	Honey, talipatra, shati, pudina satva, tulsi, mulethi, banapsha, pipali, vasaka
Dabur active blood purifier	Manjistha, shveta sariva, khadir, nimba, guduchi, honey
Dabur Honitus cough drops tab	Shudh madhu, pudina satva, nilgiri tel, sunthi tel, kapoor
Dabur baby janma ghunti	Anjeer, draksha, vidanga, svarnapatri, vacha, palash beej, aragvadha, unnab, honey, mishereya satva, shatapatrika ark
Dabur mensta syrup	Ashoka, dhatakipushpa, nisoln, kalaunji, ajwain, sunthi, amla, baheda, hareetaki, mukta, safed jeera, vasaka, pipali, ghritkumari, daruharidra, rakt chandan, amarasthi, neel kamal, hing, honey

Table 1.1 (continued)

Product	Description
Himani sardi jaa cough syrup	Tulsi, bibhitaki, haridra, sunthi, marich, pipali, mool, vach, kantakari, yasti madhu, kasmard, bharangi, kayaphala, navsadar, lavanga, pudina ka phool, kapoor, honey, chyawan concentrate, surasar
Vicks cough drops tab	Karpoor, pudina ka phool, nilgiri tel, honey
Strepsils honey and lemon tab	2,4 dochlorobenzyl alcohol, amylmetacreat BP, honey
Amartam madhu panchamrit	Madhu yasti, tulsi, brahmi, pan ka rash, madhu sita
Chirayu madhuras	Madhu, tulsi ras, adrak ras, sita
Madhushala amrit	Kalmegh, hansapag, bhringraj, ajwain, hing, pudina, adrak, sonapatra, triphala, carica papaya, anatmool, jau, biranga, karchi, dhania, pit papda, gulaucha, ashshawra, yasti madhu, grit kumari, tepurosia pupurea, bhujamalki
Ripanto ointment	Oleum lini, oleum sesame, bees wax

(as per taste) and one spoon of honey in each cup and mix and enjoy delicious and beneficial tea. It can also be consumed after cooling it in the refrigerator.

(v) **Honey Lemon Drink**

Honey is a natural sweetener; therefore, it can be used as an alternative to artificial white sugar which is not considered health-friendly. To prepare the honey lemon drink, the juice of half a lemon may be added to a glass of water along with a pinch of spices (black salt/black pepper). It is very tasty and highly beneficial for the digestive system.

Similarly, honey jam, honey candy, honey fruit shakes, and many other products may be prepared at home without much expenditure. Some of the value-added products of honey are tabulated in Table 1.1.

1.11 Conclusion

Honey is one of the most precious gifts of nature. Its nutritional, cosmeceutical, and therapeutic effects have been proven for ages. There is no other natural sweetener with such magnificent properties in this universe. The value of food ingredients increases many folds after the addition of honey. Honey proves to be highly effective as a promising therapeutic agent that can be used as an adjunct treatment for many biologically diverse diseases. Its distinctive properties and benefits make it a wonderful product among all insect products.

Chapter 2
Pharmacological Properties of Honey

Honey, a natural substance produced by bees from the nectar of flowers, has been a subject of interest due to its diverse pharmacological properties. This abstract provides a concise overview of the medicinal potential of honey, encompassing its antioxidant, anti-inflammatory, antimicrobial, and wound-healing properties. The unique composition of honey, including sugars, enzymes, vitamins, and polyphenols, contributes to its therapeutic effects. Studies have demonstrated the efficacy of honey in various medical applications, ranging from the management of chronic wounds and burns to its potential role in combating infections. Additionally, honey exhibits neuroprotective and cardiovascular benefits. Understanding the pharmacological properties of honey holds promise for its integration into healthcare practices, offering a natural and multifaceted approach to health and well-being.

2.1 Introduction

Honey, a natural sweet substance produced by honeybees through the enzymatic transformation of flower nectar, has long been cherished for its culinary appeal. Beyond its traditional use as a sweetener, honey has garnered increasing attention for its diverse pharmacological properties, positioning it as a multifaceted therapeutic agent. The intricate composition of honey, comprising sugars, enzymes, vitamins, minerals, and polyphenols, contributes to its unique medicinal potential. This introduction aims to explore and elucidate the various pharmacological facets of honey, shedding light on its antioxidant, anti-inflammatory, antimicrobial, and wound-healing attributes. Antioxidant capabilities of honey arise from its rich content of polyphenols, flavonoids, and other bioactive compounds. These constituents have been recognized for their ability to neutralize free radicals, offering protective effects against oxidative stress-related diseases. Moreover, honey's anti-inflammatory properties have been studied in the context of various inflammatory conditions, showcasing its potential as a natural remedy. The antimicrobial prowess of honey is particularly noteworthy. Studies have demonstrated its effectiveness against a spectrum of bacteria, fungi, and even some viruses. This broad-spectrum antimicrobial activity is attributed to the combined effects of its low pH, osmotic pressure, and the presence of specific compounds such as hydrogen peroxide and bee-derived peptides.

In the realm of wound healing, honey has been employed since ancient times. Its viscosity creates a protective barrier over wounds, preventing infection, while its ability to stimulate tissue regeneration accelerates the healing process. Modern research has validated honey's efficacy in managing chronic wounds, burns, and various dermatological conditions. As we delve into the pharmacological properties of honey, it becomes evident that this natural elixir extends its benefits beyond external applications. Emerging studies suggest honey's potential role in cardiovascular health, neuroprotection, and metabolic disorders (Farooqui et al. 2011). The intricate interplay of its components appears to contribute to a range of systemic effects, opening avenues for further exploration in medical research.

Therefore, pharmacological properties of honey transcend its reputation as a delightful sweetener, positioning it as a valuable resource in the realm of natural medicine. This exploration aims to provide a comprehensive overview of the current understanding of honey's medicinal potential, paving the way for continued research and integration into therapeutic practices.

2.2 Pharmacological Properties

Honey, recognized for centuries for its medicinal properties, owes its pharmacological benefits to a rich composition of various bioactive compounds. Notably, its potent antioxidant properties, including flavonoids, polyphenols, and enzymes, play a crucial role in neutralizing free radicals, preventing oxidative stress linked to chronic diseases and aging (Fig. 2.1). Furthermore, honey's well-established anti-inflammatory effects, both internal and external, make it valuable in conditions like arthritis and inflammatory disorders by modulating inflammatory pathways (Battino et al. 2021). With natural antimicrobial and antibacterial properties, honey's high sugar content, low pH, and the presence of hydrogen peroxide inhibit bacterial growth, and certain varieties like Manuka honey exhibit enhanced antibacterial effects due to components like methylglyoxal (MGO) (Mandal and Mandal 2011). Used in wound healing since ancient times, honey creates a protective barrier, promotes tissue regeneration, and accelerates healing by drawing out excess fluid from wounds. Its soothing effect on the throat, coupled with antimicrobial properties, makes honey a common ingredient in cough syrups and throat lozenges, providing relief from cough symptoms and sore throat discomfort. Additionally, honey demonstrates positive effects on the gastrointestinal system, soothing and healing issues like gastritis and ulcers, while also acting as a prebiotic, promoting the growth of beneficial gut bacteria (Abuelgasim et al. 2021). Some studies suggest potential

2.2 Pharmacological Properties

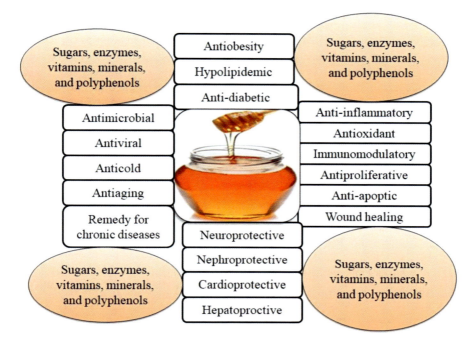

Fig. 2.1 Showing the various biological, medicinal, and protective effect of honey against various health ailments

cardiovascular benefits, linking honey to improved lipid profiles, reduced atherosclerosis risk, and better management of cardiovascular risk factors (Al-Waili et al. 2013). Certain honey types exhibit antiviral properties, showing promise in inhibiting virus replication and suggesting implications in managing viral infections. Emerging research indicates potential neuroprotective properties, as honey may protect the brain from oxidative stress and inflammation, offering benefits in neurological disorders (Fadzil et al. 2023). Considering these diverse and multifaceted pharmacological properties encompass antioxidant, anti-inflammatory, antimicrobial, and wound-healing aspects, with specific benefits varying based on honey type and botanical origin (Andreu et al. 2015).

2.2.1 Antibacterial Properties

The antibacterial property of honey is the most studied aspect of honey and is the easiest to test. Honey is acidic in nature and possesses a pH of 3.5–5.0, which means it is acidic and has high sugar content (Bhattarai 2021). This property is due to hydrogen peroxide produced during the formation of gluconic acid from glucose. The glucose oxidase enzyme is usually inactive in pure honey. Therefore, in diluted honey with the correct pH, the antibacterial activity is mostly due to hydrogen peroxide. The primary importance of such a mechanism results from the need to protect the raw honey contained in the colony until higher sugar concentrations are achieved. Honey has antibacterial properties, and honey is recognized as no infection by "stimulating the immune system" to defeat the interlopers (Cohen and Siegel 2021). Recently, a substantial description confirms that it is an innate immunity enhancer. Some recent reports describe the activity of "B60 lymphocytes and T lymphocytes" to increase the quantity (number) in a cell sample or to trigger neutrophils (Kumari et al. 2021). The acidic pH of honey significantly contributes to the digestion of dead bacteria in the phagocytic method.

2.2.2 Antimicrobial Properties

Clinically, the importance of the antimicrobial property of honey cannot be overstated, the immune responses (body) may be insufficient to clear the infection. For those microorganisms which develop resistance against antibiotics, honey has proved to be an effective antimicrobial providing resistance against both non-pathogenic and pathogenic microorganisms, i.e., yeasts, bacteria, fungi, etc. (Massoura 2020). The antimicrobial potential of honey hinders the development of bacteria or kills bacteria completely. The acidity of honey inhibits the growth of microbes (Almasaudi 2021). It is already reported that most of the therapeutic properties of honey are partly due to its plant origin. Organic solvents such as n-hexane, ethyl acetate, diethyl ether, and chloroform are carried out using liquid-phase and solid-phase extraction methods for removal (Nazir et al. 2017). It can be used to remove compounds showing their activity. The isolated compounds have been attributed to integrate flavonoids, and other compounds such as phenolic acids, proteins, lipids, ascorbic acid, carotenoids, natural acids, etc. Other important effects of honey are fructooligosaccharides or other types of oligosaccharides that have prebiotic properties. Honey has been used as a traditional medicine for bacterial infections since historical times. Studies have shown that honey, which contains *Enterobacter aerogenes*, *E. coli* (*Escherichia coli*), *S. aureus*, and *S. Typhimurium* (*Salmonella typhimurium*) has been the best preventive measure against human pathogens (Lusby et al. 2005). Honey is effective against methicillin-resistant *S. aureus*, vancomycin-resistant enterococci and streptococci (Rani et al. 2017). Honey has been proven for its antimicrobial properties for centuries, and people all over the world are using it for various infections and wounds (Kumar et al. 2010a, b). The following factors are responsible for the antimicrobial potential of honey:

High Sugar Content: Honey has low water content and high sugar content, primarily in the form of glucose and fructose. This high sugar concentration creates an environment where bacteria and other microorganisms cannot easily thrive. The sugar content draws moisture from the environment, creating a hypertonic solution that dehydrates and kills many microorganisms (Ariyamuthu et al. 2022).

Low pH: Honey is acidic, with a pH ranging from 3.2 to 4.5. This acidic environment inhibits the growth and survival of many bacteria, as they prefer a more neutral pH.

Hydrogen Peroxide Production: Honey contains glucose oxidase, an enzyme that produces hydrogen peroxide when honey comes into contact with moisture. Hydrogen peroxide has antiseptic properties and can help kill bacteria.

Phytochemicals: Honey contains various phytochemicals, including flavonoids and phenolic compounds that have antioxidant and antimicrobial properties. These compounds can help prevent the growth of bacteria and fungi.

Less Water Activity: The water activity of honey is typically low, which means that there is very less amount of water available for microorganisms to grow (Olaitan et al. 2007). This is another factor that contributes to its long shelf life and antimicrobial properties.

High Osmolarity: Honey's high osmolarity (concentration of solute particles) causes osmosis in the bacterial cells leading to the release of water, causing them to shrink and die.

Producing Inhibitory Substances: Bees also add substances to honey during the honey-making process that can inhibit the growth of microorganisms.

While honey does possess these natural antimicrobial properties, however, extent of these properties varies among different types of honey. The variation may depend upon the floral source, geographic location, and processing methods. Manuka honey, for example, has potential antimicrobial properties due to a compound called methylglyoxal (MGO) (Johnston et al. 2018). Due to the antimicrobial nature of honey, it is used to treat wounds, sore throats, and minor infections for its antimicrobial properties (Kumar et al. 2010a, b. However, it is essential to use caution and consult with a healthcare professional when considering honey as a cure/management for medical conditions. In some cases, medical-grade honey may be recommended for wound care.

2.2.3 Immunomodulatory Properties

In immune response, B cells, T cells, cytokines, growth factors, TNF-α, IL-1β, and IL-6 play important roles. The honey has shown the mitogenic activities on B cells and T cells. The previous findings reveal that natural honey can induce the proliferation of immune cells and release of TNF-α, IL-1β, and IL-6 (Samarghandian et al. 2017). Honey is found to be effective in wound healing. Honey is a traditional medicine and natural dietary supplement. Recently, it has attracted more consideration in the treatment of some diseases (Eteraf-Oskouei and Najafi 2013). Honey proves beneficial in oral mucositis and induces apoptosis in prostate cancer cells. Flavonoids and phenolics of honey are used to block the cell cycle. It has been revealed by previous studies that consumption of honey increases the production of antibodies in the immune responses (Hegazi et al. 2013).

The formation of SCFAs (short-chain fatty acids) may hinder the ingestion of honey, and SCFAs are also accountable for immunomodulatory activities. Honey sugars are able to influence the immune response. Furthermore, there are no sugar components that are principally accountable for the immunomodulation activities. The conclusion is that honey also has anti-immunomodulatory and anti-inflammatory activities. Honey modulates cytokine production in cells. Honey contributes to skin healing by boosting the skin immune system to fight infection, promoting wound healing in the skin (Scepankova et al. 2021a, b). Researchers has stimulated the production of the cytokine tumor necrosis factor-alpha (TNF-α) by the tumor cell (Masad et al. 2021).

The immunomodulatory effect of honey is mainly due to polyphenolic compounds. Polyphenols have multiple phenolic groups associated with complex structures. Phenolic composition of honey depends on its floral source. Polyphenol quercetin was found to inhibit the release of TNF-α, IL-1β by macrophages. In different autoimmune diseases, immunomodulatory effect of quercetin is noticed (Talebi et al. 2020). Polyphenol luteolin was found to inhibit gene expression of pro-inflammatory cytokines, moreover, proliferation of auto reactive T cells is also inhibited by luteolin (El-Seedi et al. 2022).

2.2.4 Antioxidant Properties

Honey, celebrated for its delectable sweetness, is also endowed with a diverse array of antioxidants that contribute significantly to its potential health benefits (Fig. 2.2). Antioxidants play a crucial role in safeguarding our cells from the detrimental effects of free radicals, by-products of various bodily processes that can lead to oxidative stress—a factor implicated in aging, chronic diseases, and inflammation (Lobo et al. 2010). The rich composition of honey encompasses various antioxidants, including phenolic compounds, flavonoids, and enzymes such as catalase and glucose oxidase (Luchese et al. 2017). Working in harmony, these antioxidants actively neutralize free radicals, mitigating oxidative stress within the body. Phenolic compounds found in honey, such as caffeic acid and quercetin, exhibit potent antioxidant and anti-inflammatory properties (Biluca et al. 2020). A distinctive aspect of honey's antioxidant activity is its capacity to directly scavenge free radicals, preventing cellular damage. Moreover, research indicates that honey can augment the body's intrinsic antioxidant defense mechanisms. Consumption of honey has been linked to increased levels of key antioxidant enzymes like superoxide dismutase (SOD) and glutathione peroxidase (GPx) (Erejuwa et al. 2012a, b).

The floral source of honey significantly influences its antioxidant composition. Varieties derived from different plants exhibit varying antioxidant levels, with darker honey, like buckwheat honey, often containing higher antioxidant content than lighter varieties such as clover honey (Becerril-Sánchez et al. 2021). Additionally, processing and storage conditions for honey can impact its antioxidant properties, with raw and minimally processed varieties retaining more natural

2.2 Pharmacological Properties

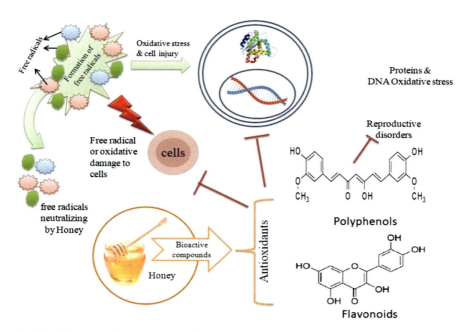

Fig. 2.2 Showing action mechanism of honey against various irregulated physiological functions

antioxidants compared to highly processed alternatives (Scepankova et al. 2021a, b). Beyond countering oxidative stress, the antioxidants in honey are believed to contribute to its potential health-promoting effects. It has been suggested that honey may support cardiovascular health, aid in wound healing, and even exhibit anticancer properties in laboratory studies (Scepankova et al. 2017).

Scientists, exploring ways to counteract the harmful effects of free radicals and reactive oxygen species (ROS) in aging and disease processes, are investigating honey's potential as a multifaceted antioxidant (Qadir et al. 2020). Various methods, including DPPH, FRAP, ORAC, ABTS, and TEAC assays, are employed to measure honey's antioxidant activity (Gorjanović et al. 2013). Notably, the evaluation of antioxidant activity should utilize rigorously tested and standardized methods. The capacity of honey to reduce oxidative reactions and scavenge free radicals within organisms is attributed to its bioactive antioxidant compounds, including flavonoids and phenolic (Cianciosi et al. 2018). Flavonoids, a class of naturally active compounds, neutralize reactive oxygen species (ROS) generated during metabolism, preventing tissue damage and showcasing honey's anti-inflammatory properties. The antioxidant properties of honey, primarily attributed to its phenolic content, are crucial in countering oxidative stress implicated in various diseases, such as atherosclerosis, cancer, and aging (Cianciosi et al. 2018). The concentration of these antioxidants, including ascorbic acid, tocopherol, catalase, and polyphenols, contributes to honey's robust antioxidant activities, with darker varieties exhibiting higher values than lighter ones. Therefore, honey emerges as a multifaceted natural source of antioxidants, effectively combating free radicals, fortifying the body's own defenses, and potentially offering a spectrum of health benefits. Its intricate antioxidant profile combined with its sweet and versatile nature, positions honey not only as a delightful culinary delight but also as a potential ally in promoting overall well-being.

2.2.5 Inflammatory Properties

Inflammation, a defensive biological response to pathogens, is an innate immune reaction that can lead to oxidative stress. Honey, recognized as an immune-modulatory agent, exhibits its anti-inflammatory effects through multiple mechanisms (Fig. 2.2). Primarily honey stimulates the production of inflammatory mediators such as cyclooxygenase-2 and prostaglandin E2, while concurrently down-regulating inflammatory transcription factors (Martinez-Armenta et al. 2021). Notably, honey reduces the activity of both cyclooxygenase-1 and cyclooxygenase-2, ultimately lowering the concentrations of prostaglandins E2 and F2α in human blood plasma.

The inflammatory response can manifest as either acute or chronic. Acute inflammation is characterized by itching, pain, and reddishness, serving as indications of the body's natural defense against pathogens (Bellamy et al. 2020). In contrast, chronic inflammation, marked by prolonged symptoms due to inadequate treatment, can damage tissues and hinder the healing process.

Anti-inflammatory properties of honey stem from its phenolic content, which specifically suppresses the pro-inflammatory activities of cyclooxygenase-2 (Ranneh et al. 2021). In essence, honey acts as a natural remedy, mitigating excessive inflammation by addressing the root causes and promoting a balanced immune response. In practical terms, when the body initiates an inflammatory response, it generates free radicals that can exacerbate the situation. Diverse components of honey, including antioxidants and enzymes, collaborate to neutralize these free radicals, preventing further damage (Ahmed et al. 2018). Furthermore, honey's distinct compounds directly soothe inflamed areas, analogous to a cooling spray on a metaphorical fire of inflammation. Notably, different types of honey, derived from various flowers, exhibit unique anti-inflammatory properties. Darker honey, like buckwheat honey, stands out for its heightened efficacy in reducing inflammation, attributed to its richer concentration of beneficial compounds (Farooqui et al. 2011). However, it is crucial to recognize that honey's processing methods impact its anti-inflammatory potential. Raw or minimally processed honey retains more of its natural goodness, unlike heavily processed variants that may lose some of their anti-inflammatory powers.

Therefore, honey extends beyond its role as a sweetener, emerging as a natural ally against excessive inflammation. Whether enjoyed on its own or added to tea, honey's antioxidants combat free radicals, while its distinctive compounds directly alleviate inflammation. By choosing minimally processed honey, individuals can maximize its anti-inflammatory benefits and contribute to maintaining a harmonious immune response in the body.

2.2.6 Prebiotic Properties

Prebiotics are indigestible components in food that support the development of probiotic microbiota in the gut, fostering overall host health by modulating microbial balance (Dahiya and Nigam 2022). They have the ability to impede the growth and activity of undesirable microflora. Honey, due to its saccharides, serves as potential prebiotic, offering fermentable carbohydrates to selective gut microbiota and promoting the prebiotic effect by generating beneficial metabolites. The term "prebiotics" was coined by Gibson and Roberfroid in 1995, describing substances that lack caloric value yet exhibit characteristics such as resistance to digestive enzymes and low pH, stimulation of specific gut microbes, and fermentation by intestinal microbiota.

Prebiotic potential of honey arises from its oligosaccharides, which encourage the growth of bifidobacteria and lactobacilli (Mustar and Ibrahim 2022). These prebiotics benefit the host by favoring the survival of beneficial bacteria, such as Lactobacillus and Bifidobacterium, while selectively inhibiting the growth of undesirable microorganisms like Helicobacter pylori, Salmonella species, and Clostridium. Meanwhile, probiotics, live microorganisms with positive health effects, have been shown to improve conditions like inflammatory bowel diseases, obesity, and diabetes. Probiotics exhibit antimicrobial, anti-inflammatory, and anticarcinogenic properties. In the context of honey as a prebiotic, it contains fructooligosaccharides (FOS) and inulin, both polysaccharides with β-glycosidic linkages that resist hydrolysis by human gut enzymes (Brandelli 2021). These compounds, commonly found in honey produced by Apis mellifera, contribute to a healthy gut microbiome by serving as prebiotics. Notably, inulin and FOS, prevalent in foods like spinach, peas, asparagus, lentils, oats, onion, garlic, and chicory, maintain gut microflora health. A robust gut microbiota is crucial for preventing common health issues such as heart diseases, diabetes, and inflammatory bowel diseases, emphasizing the importance of bifidobacteria for optimal gut function.

2.2.7 Antifungal Properties

Honey is recognized for its efficacy against various fungal infections such as ringworms, nail fungus, and athlete's foot, attributing its antifungal properties to aromatic acids present in its composition. Key aromatic acids include methyl cinnamate, cinnamoylglycine, benzyl cinnamate, and terpenoids. The antifungal prowess of honey arises from its acidity, high osmolarity, hydrogen peroxide content, and diverse phytochemicals, collectively inhibiting the growth or killing the fungus.

One of honey's unique features contributing to its antifungal capabilities is its low water content (around 17–18%), resulting in low water activity that creates an inhospitable environment for fungal growth by reducing available moisture (Maddocks and Jenkins 2013). Additionally, honey's high sugar content, primarily composed of glucose and fructose, establishes a hypertonic environment, drawing water out of fungal cells through osmosis and causing them to shrink and perish. The natural acidity of honey, typically ranging from pH 3.2 to 4.5, disrupts fungal cell membranes and metabolic processes, impeding their growth (Israili 2014). Furthermore, the enzyme glucose oxidase in honey produces hydrogen peroxide when in contact with moisture, providing additional antifungal properties that aid in eliminating fungal spores and cells.

Repertoire phytochemicals of honey, including flavonoids and phenolic compounds, possess antioxidant and antimicrobial properties, contributing to the inhibition of fungal growth. Bee-derived compounds added during honey production also play a role in inhibiting microorganisms.

Studies, such as those by Mekky in 2007, have demonstrated honey's effectiveness in inhibiting the growth of specific fungi, such as *Aspergillus flavus*, and reducing aflatoxin B1 and B2 levels (Sinha et al. 2023). Moreover, the antifungal activity of honey extends to combating yeast species like Candida albicans, as well as molds like Penicillium chrysogenum and Aspergillus baumannii, as reported by (Sinha et al. 2023). Honey's longstanding reputation for antimicrobial properties, including antifungal attributes, positions it as a valuable natural remedy for fungal infections like athlete's foot, nail fungus, and yeast infections (Mandal and Mandal 2011). Whether applied topically or ingested, honey's multifaceted antifungal mechanisms underscore its potential in managing such infections. However, it's essential to recognize that the efficacy of honey may vary based on factors such as honey type, source, and the specific fungal strain involved. For persistent or severe fungal infections, consulting with a healthcare professional for accurate diagnosis and treatment is crucial, combining medical-grade antifungal medications with natural remedies like honey when appropriate.

2.2.8 *Antiviral Properties*

Honey, a natural sweet substance produced by bees from flower nectar, has a rich history in traditional medicine for addressing various ailments, including those caused by viruses. Its therapeutic potential extends to antiviral properties attributed to its phytochemical composition (Mackin et al. 2023). The antiviral characteristics of honey stem from its high acidity, osmotic pressure, and hydrogen peroxide concentration.

Hydrogen Peroxide Production: An enzyme in honey, glucose oxidase, facilitates the production of hydrogen peroxide when honey encounters moisture. This antiviral agent plays a pivotal role in inactivating viruses.

High Sugar Content: Predominantly composed of glucose and fructose, honey's high sugar concentration creates a hypertonic environment, leading to the dehydration and inactivation of viral particles.

Acidity: With a pH ranging from 3.2 to 4.5, honey's natural acidity disrupts the structure and function of viral particles, diminishing their infectious potential.

Phytochemicals: Honey's diverse phytochemicals, including flavonoids and phenolic compounds, possess antioxidant and antimicrobial properties, demonstrating antiviral effects in laboratory studies by interfering with viral replication.

Immune System Modulation: Some studies suggest that honey can modulate the immune system by stimulating the production of cytokines and other immune factors, enhancing the body's ability to combat viral infections (Al-Hatamleh et al. 2020). Honey shows evidence of interfering with the attachment of viruses to host cells, hindering the initial stages of infection. Anti-inflammatory properties of honey may alleviate the excessive inflammatory response seen in viral infections, indirectly supporting the body's antiviral defenses (Al-Hatamleh et al. 2020). While research indicates promising antiviral properties of honey, its efficacy varies among different viruses, necessitating further study, including clinical trials, to determine its specific effectiveness. Honey is commonly used to soothe symptoms of viral infections, providing relief from sore throats and coughs (Abuelgasim et al. 2021). However, it should not be considered a sole treatment for serious viral infections, and medical intervention is crucial in such cases.

Caution is advised, especially regarding infants under 1-year-old, as honey poses a risk of infant botulism. Individuals with bee product allergies should use honey cautiously. Despite its antiviral properties, honey is a complementary element in health and wellness, not a substitute for conventional medical treatments in the case of severe viral infections. Ongoing research will enhance our understanding of honey's potential as an antiviral therapy against diverse viruses.

2.2.9 Anticarcinogenic Properties

Honey, a natural sweet substance produced by bees from flower nectar, has garnered attention for its potential anticarcinogenic properties. Studies suggest that honey may contribute to cancer prevention and treatment through various mechanisms. Its phenolic content has been found to activate caspase-3 in cancerous cells, inducing apoptosis by regulating the expression of proteins P53, Bax, and caspase 3, while also modulating antiapoptotic proteins (Bcl-2) (Talebi et al. 2020). Furthermore, honey exhibits a chemo-preventive effect against cancer by regulating oxidative stress and demonstrating strong activity against mutagenicity linked to carcinogenicity. Studies reveal honey's ability to block the cell cycle in the G0/G1 phase in colon cancer cell lines, indicating its role in suppressing abnormal cell division (Subramanian et al. 2016). The anticancer effects of honey encompass inducing apoptosis, antimutagenic activity, interference with cell signaling pathways, and antiproliferative activity. Its impact on cell cycle progression and depolarization of mitochondrial membranes in cancerous cells positions honey as a natural anticancer agent. The G1/S phase transition regulation in animals, crucial for cell division, is reported to be influenced by honey, showcasing its potential in arresting the cell cycle. Additionally, antiproliferative activity of honey in the G0/G1 phase has been observed in various cancer cell lines, including glioma, colon, and melanoma (Mumtaz et al. 2020). While ongoing research explores honey's anticarcinogenic potential, it is essential to recognize its variable properties based on factors such as type, processing methods, and quality. Despite its promise in a cancer-preventive diet, honey should not be viewed as a standalone cancer treatment. Given the complexity of cancer and its multifaceted risk factors, individuals facing or at risk of cancer should consult healthcare professionals specializing in oncology for comprehensive evaluation and treatment planning.

2.2.10 Immunity Boosting Activity

Honey is recognized for its reported benefits in various diseases, including cough, upper respiratory infections, allergies, common colds, asthma, hay fever, sinusitis, ulcers, gingivitis, inflammation, diarrhea, infections, conjunctivitis, constipation, and colitis. Its effectiveness is attributed to its ability to increase B and T lymphocytes, antibodies, and natural killer (NK) cells, thereby stimulating the immune system in both rats and cell cultures. Honey's unique composition triggers the release of cytokines such as IF-1, IF-6, and TNF-α, inducing an immune response (Ahmed et al. 2018). The presence of glucose in honey contributes to a respiratory burst, generating hydrogen peroxide with bacteria-killing activity. The glucose in honey serves as the primary substrate for glycolysis in macrophages, enabling them to function effectively in damaged tissue. Honey is frequently lauded for its potential to support the immune system due to its bioactive compounds and traditional medicinal use. While it cannot substitute a healthy lifestyle and balanced diet, honey's antioxidant properties, anti-inflammatory effects, antimicrobial action, soothing properties, immune system modulation, gut health promotion, wound-healing aid, and provision of quick energy make it a valuable addition to overall immune support (Eteraf-Oskouei and Najafi 2013). It is crucial to consider the quality of honey, with raw and unprocessed varieties generally deemed more potent. While honey can contribute to immune health, a comprehensive approach involves a nutritious diet, regular exercise, sufficient sleep, stress management, hygiene practices, and consultation with healthcare professionals for personalized guidance, especially in the case of specific health concerns or conditions (Samarghandian et al. 2011, 2017).

2.2 Pharmacological Properties

2.2.11 Antidiabetic

Diabetes mellitus, encompassing Type-1 and more prevalent Type-2 diabetes, remains a major global health concern despite various medical interventions. In the pursuit of alternative treatments, medical science is exploring natural products such as honey, a traditional folk medicine dating back to the beginning of human civilization. Comprising fructose, glucose, and various complex sugars, honey also contains organic acids, minerals, proteins, amino acids, enzymes, and polyphenols (Pavlova et al. 2018). The fructose/glucose ratio in honey, coupled with its lower glycemic index compared to glucose, is thought to contribute to its potential hypoglycemic effects (Zamanian and Azizi-Soleiman 2020; Erejuwa et al. 2012a, b). Fructose stimulates liver cell enzyme glucokinase, promoting glycogenesis, while the antioxidant properties of honey may protect the pancreas from oxidative stress (Pasupuleti et al. 2020). However, caution is advised in interpreting honey as a diabetes treatment, considering its impact on blood sugar levels. Despite being viewed as a healthier alternative to refined sugar, honey's moderate to high glycemic index can lead to rapid spikes in blood sugar. For individuals with diabetes or those at risk, managing carbohydrate intake, practicing portion control, and diligent monitoring of blood glucose are essential. While some studies suggest a modest positive effect on insulin sensitivity, honey is not considered an antidiabetic or recommended treatment for diabetes. It may be cautiously incorporated into a well-balanced diet in small amounts, but overall, individuals with diabetes should approach honey consumption under healthcare professional guidance. The primary focus for diabetes management remains blood glucose monitoring, adherence to a diabetes-friendly diet, and careful management of carbohydrate intake.

2.2.12 Antihypertensive

Hypertension can lead to deformities in coronary vasculature and myocardial structure, ultimately resulting in congestive heart failure and heart dysfunction. Recent findings suggest that honey intake may reduce the risks associated with hypertension. Studies using honey derivatives, such as a honey-based beverage containing γ-aminobutyric acid (GABA), have shown beneficial effects on hypertension in rats, with a significant reduction in systolic blood pressure observed after 7 weeks of supplementation. Research by Erejuwa et al. (2011) has highlighted the potential antihypertensive effects of honey feeding on hypertensive rats, demonstrating a significant reduction in systolic blood pressure and an inhibition of angiotensin-I-converting enzyme (ACE), a key player in the hypertensive mechanism. Honey's antioxidant properties, vasodilatory effects, anti-inflammatory properties, potassium content, stimulation of nitric oxide production, and stress-reducing qualities contribute to its potential antihypertensive effects. However, it is crucial to understand that while honey may offer potential benefits for blood pressure control, it should not be considered a standalone treatment for hypertension. Managing high blood pressure requires a comprehensive approach, including dietary changes, regular physical activity, and, in some cases, medication prescribed by a healthcare professional. Individuals with hypertension or concerns about blood pressure should consult with a healthcare provider for personalized evaluation and guidance on managing their condition. While honey can be a part of a heart-healthy diet, its role in blood pressure management should be discussed with healthcare professionals to ensure a well-rounded and effective treatment plan.

2.2.13 Cardio-Protective Role of Honey

The cardio-protective role of honey spans a spectrum of advantageous effects that collectively contribute to cardiovascular health. Notably, honey has been associated with improvements in lipid profiles, demonstrating reduced levels of total cholesterol, low-density lipoprotein (LDL) cholesterol, and triglycerides critical for sustaining cardiovascular well-being and mitigating the risk of atherosclerosis (Battino et al. 2019). Its richness in antioxidants, such as flavonoids and polyphenols, plays a pivotal role in safeguarding the cardiovascular system by neutralizing free radicals, thus preventing the oxidative stress implicated in the development of cardiovascular diseases (Ghagane and Akbar 2023). Complementing this, honey's anti-inflammatory properties address chronic inflammation, a key contributor to cardiovascular disease progression, reducing the risk of conditions like atherosclerosis and fostering overall heart health (Chepulis 2008). Studies suggest that regular honey consumption may modestly regulate blood pressure, exerting a hypotensive effect that aids in lowering elevated blood pressure and thereby reducing the risk of complications related to hypertension (Khalil and Sulaiman 2010). Furthermore, honey may contribute to the improvement of endothelial function, crucial for regulating vascular tone and maintaining blood vessel health, ultimately associated with a diminished risk of cardiovascular events (Hashim et al. 2021). The combined effects of honey's antioxidant, anti-inflammatory, and lipid-lowering properties position it as a potentially effective agent in reducing the risk of atherosclerosis a condition marked by arterial plaque build-up. Moreover, honey has been studied for its impact on various cardiovascular risk factors, including obesity and diabetes, indirectly contributing to the prevention of heart-related complications. The vasodilatory effects of honey, coupled with its capacity to enhance endothelial function, promote better vascular health, leading to improved blood flow and reduced strain on the heart. While more research is needed, some studies suggest that honey's cardio-protective properties may extend to the prevention of myocardial infarction, with its anti-inflammatory and antioxidant effects potentially shielding the heart muscle from damage (Din et al. 2023). It is crucial to note that the cardio-protective benefits of honey may vary based on factors like honey type, purity, and individual lifestyle. While honey can be a valuable component of a heart-healthy diet, moderation is key within a balanced lifestyle that includes regular exercise and a well-rounded diet. Individuals with pre-existing heart conditions should seek personalized advice from healthcare professionals.

2.2.14 Wound-Healing Properties of Honey

The wound-healing properties of honey have been acknowledged since ancient times, and contemporary research continues to reveal its multifaceted aspects that render honey a valuable agent in tissue repair (Nikhat and Fazil 2022). Honey's remarkable capabilities in wound healing can be attributed to several key factors. Firstly, it creates a protective barrier over wounds, safeguarding them from external contaminants and pathogens, thereby minimizing infection risks and providing an optimal environment for the healing process (Kaiser et al. 2021). The acidic pH and low water content of honey contribute to its natural antimicrobial properties, inhibiting the growth of bacteria and microorganisms at the wound site (Israili 2014). The high sugar content induces an osmotic effect, drawing excess fluid out of the wound, reducing swelling, and creating a conducive environment for cellular repair. Enzymes in honey, such as glucose oxidase, gradually release hydrogen peroxide, providing additional antimicrobial support to prevent and eliminate bacterial infections in the wound (Aurongzeb and Azim 2011). Importantly, honey stimulates the immune response at the wound site, expediting debris and dead tissue clearance, fostering a clean and conducive milieu for healing. The rich content of bioactive compounds, including flavonoids and polyphenols, contributes to honey's antioxidant properties, playing a pivotal role in neutralizing free radicals that could hinder the healing process and supporting tissue regeneration. Impact of honey extends beyond superficial wounds, proving effective in chronic and hard-to-heal wounds. Its ability to stimulate angiogenesis, the formation of new blood vessels, enhances blood supply to the wound area, facilitating nutrient delivery and waste removal crucial for tissue regeneration. Specific types of honey, such as Manuka honey, exhibit unique properties attributed to components like methylglyoxal (MGO), further enhancing their wound-healing potential with additional antimicrobial effects, making them particularly effective in managing infections in challenging wounds. In conclusion, honey's wound-healing aspects encompass a comprehensive array of mechanisms, including the formation of a protective barrier, antimicrobial properties, osmotic effects, enzymatic activity, immune modulation, antioxidant effects, and the promotion of angiogenesis. These properties collectively position honey as a natural and versatile adjunct in wound care, offering benefits across a spectrum of wound types and aiding in the overall process of tissue repair. It is essential to note that while honey is generally considered safe for topical use, consultation with healthcare professionals is advised, especially for individuals with allergies or specific medical conditions.

Chapter 3
Herbs as Therapeutics and Healers

3.1 Introduction

People have relied on plants for their health care since pre-historic times. Herbal medicines are the naturally occurring substances which are derived from the plants and used to cure ailments based on indigenous practices. The term "herb" originated from the Latin word "herba" which means "a green plant." Originally, the term was generally applied to grasses, green crops, and other leafy plants. But now any plant, tree, shrub, or herb that has having economic significance and used for its medicinal, culinary, or decorative qualities is regarded as herb. Herbs can belong to various genera as well as species, and they differ in their botanical traits. Although the composition of each type of plant varies, they all include active ingredients that might influence different bodily processes. Apart from their culinary use, several herbs possess the ability to treat and avoid illnesses. As the chemical makeup of herbs changes depending on several things, their utilization varies from person to person. Some individuals utilize herbal extracts as conventional medicine, whereas others use them in the form of energizing decoctions.

Herbs are classified as annuals, biennials, or perennials depending upon their life cycle. Annuals are initiated from seeds, grow and then perish in a single year or season. Plants classified as biennials which need 2 years to complete their life cycle. Plants that thrive for longer than 2 years are called perennials. Certain herbs like dandelion are regarded as weeds (unwanted plant). Although, with the right application, a plant that is usually considered a weed can have medicinal properties.

Certain plants that belong to the vegetable category can also be categorized as herbs. For example, one of the most well-known vegetables that is also herb is garlic. Garlic not only gives meals an aromatic flavor, but it also has blood-purifying, cholesterol-lowering, and antibacterial qualities. Celery is an additional vegetable with a lengthy history of use as herbal remedy. A traditional remedy for uric acid accumulation in the body which aggravates gout and arthritis patients by causing inflammation and pain is to utilize celery seeds. Celery seeds also encourage relaxation.

3.2 Traditional Use of Herbs

Traditional uses of herbs are generally considered as safe with negligible side effects. Common herbs or their parts are used by local people as powder, juice, poultice, decoctions, and tablets either singly or in combinations. The etiology of diseases and therapeutic approaches are based on philosophical, socio-cultural, and theological consideration, taking into account patient's social and mental behavior.

Traditional medicine has a very long and vast history. It is the culmination of all the theories, beliefs, and experiences from all cultures and eras that utilized the herbal medicines to maintain health including sickness prevention, diagnosis, improvement, and treatment. People in many parts of the world still rely upon traditional medicinal systems, folklore, religious doctrine, and other mystical ideas. Presently, efforts are being made for promoting clinical research techniques to evaluate the efficacy of traditional medicines.

Herbal folklore traditions for the medicinal applications of plants were transmitted from generation to generation through familial oral histories and experiential learning before written history. Since pre-historic times, every family, tribe, and civilization had their traditional herbal treatments derived from the therapeutic plants that could be found in the area. For example, native American women make tea from the cotton plant's root to ease labor-related discomforts. In medicine, the cotton plant's root helps to constrict the uterus and promote regular menstruation following childbirth (Vincent 2011).

Cultures have historically relied on the knowledge of earlier generations to discover how to employ medicinal plants for healing. Many societies all around the world have examples of the important role that herbs have played throughout history. The first acknowledged usage of plants as medicinal agents was imprinted in the radiocarbon-dated Lascaux cave paintings in France which dates back to 13,000–15,000 BCE. On a Sumerian clay slab from Nagpur, the earliest known written record of the use of medicinal herbs to prepare medications has been found which dates back to about 5000 years. It included 12 drug preparation formulas based on more than 250 different plant species.

The oldest record of the utilization of plants in traditional medicines dates back to 2600 BCE referring 1000 herbal products in Mesopotamia. Egyptian system of medicine (about 2900 BCE) is however claimed to be the ancient practice of traditional medicine. Records of Ebers Papyrus dates back to 1500 BCE describing about 700 herb-based remedies. Wu Shi Er Bing Fang of Chinese traditional medicine dates back to 1100 BCE having 52 records of herbal medicines. Shen Nong's Materia Medica, one of the ancient Chinese herbals, contains 365 curative treatments, the most of which are plant-based extracts with a small number of mineral and animal extracts. The Greek physician Dioscorides listed over 400 plants in his writings from the first century AD. Approximately 5800 plants are listed in the Chinese Materia Medica, 2500 are known to exist in India, at least 800 are regularly collected from Africa's tropical forests, nearly 300 are currently outlined for the medical profession in Germany and thousands more are only known to traditional healers in the world's more remote corners. Records of Indian traditional medicine date back to around 5000 years back. Ayurveda is regarded as one of the oldest of the traditional medicinal systems throughout the world. About 1000 BCE ago, Charaka Samhita and Sushruta Samhita included the description of 341 and 395 plant-based drugs, respectively.

In his remarkable works "De Causis Plantarium" and "De Historia Plantarium," Theophrastus (371-287 BC) established botanical science. He categorized over 500 known medicinal herbs with detailed descriptions in these books. Hippocrates (459–370 BC) documented 300 medicinal plants and categorized according to their physiological action. Garlic was used to combat intestinal parasites; henbane, opium, mandrake, and deadly nightshade were used as narcotics; fragrant hellebore and haselwort were used as emetics; asparagus, celery, sea onion, garlic, and parsley were used as diuretics. In the eighteenth century, Linnaeus (1707–1788) gave a concise account and categorization of the species in his publication Species Plantarium (1753). The understanding and application of medicinal herbs underwent a great change in the early nineteenth century. Scientific as well as pharmacological validation began with the discovery and separation of alkaloids from many plants including poppy (1806), ipecacuanha (1817), strychnos (1817), quinine (1820), pomegranate (1878), and many others. This was followed by the isolation of glycosides. As chemical processes improved, more active components of medicinal plants were isolated including vitamins, hormones, tannins, saponosides, etheric oils, and more. The therapeutic qualities of specific medicinal plants were discovered, recorded, and passed down to the next generations in every era and every century since the emergence of ancient civilizations. The advantages of one society were transferred to another, which improved the previous properties and revealed the new ones. Today's sophisticated and advanced methods of processing and using medicinal herbs are a result of people's ongoing and unwavering interest in them (Mosihuzzaman et al. 2012; Petrovska 2012; Ody 2017; Adhikari and Paul 2018).

3.3 Herbs as Therapeutics

Herbs have long been practiced for their therapeutic and medical properties in addition to their aesthetic and culinary benefits. Since ancient times, people have looked to nature for healing from illnesses. In fact, the usage of herbal remedies in nutritional supplements, energy drinks, multivitamins, massage, and weight loss has been more and more common in recent years. The area of herbal medicine has expanded and gained credibility as a result of these applications. Additionally, the dental profession has started to utilize the therapeutic qualities of herbs to treat canker sores, gum inflammation, and tooth discomfort. Plants are the source of antiseptics, antibacterial, antimicrobial, antifungal, antioxidant, antiviral, and analgesic compounds.

Herbal medicines provide a very basic kind of therapy. Indeed, it's been recognized as the "Art of Simpling" for a very long time. Because a single herb could be used to cure a wide range of ailments, herbs were classified as "simples." Utilizing local herbs is the first principle of simpling. The kind of disease which is to be diminished in a given location is influenced by the local environment to some extent. For instance, bronchitis is more common in northern climates while parasitic diseases are more common in southern climates. In a similar vein, the herbs that grow there adopt the traits of their surroundings and are especially helpful in treating illnesses related to the local climate and other factors. There are plenty of key herbs in every location that are useful for treating the majority of ailments that arise there. Using gentle herbs is the second principle. Gentle herbs are safe to consume and have a universal impact on all bodily systems, promoting healing. The third rule is that you need to use a lot of these gentle herbs. The herb is quite mild, thus its ability to cure most illnesses will only come from high doses (Tierra 1998).

Herbs are incredibly adaptable plants. As the herb contains multiple components, each with unique medicinal capabilities, it can be used to cure multiple conditions. For example, glucoside silicon, a naturally occurring molecule, makes white willow bark a potent pain reliever. Aspirin was initially synthesized chemically as glucoside silicon. Nevertheless, willow is also well-known for being a sedative and an extremely potent disinfectant (Miner and Hoffhines 2007). Herbs can heal in a variety of ways possessing various unique therapeutic properties.

3.4 Antioxidant Properties

Herbs have some antioxidant properties that are beneficial to human health. Antioxidants are thought to postpone and stop the harm that free radicals may cause to the cells. When cells use the vital oxygen they require, harmful by-products known as free radicals are produced. There are several different ways that free radicals might arise. For instance, free radical generation is accelerated by exposure to external factors such as pesticides, alcohol, narcotics, and pollution. The formation of free radicals can also be increased by radiation, poor diets, and extended sun exposure. *Lawsonia inermis, Ocimum sanctum, Cichorium intybus, Piper cubeba, Punica granatum, Allium sativum, Delonix regia, Terminalia chebula, Terminalia bellirica, Mangifera indica, Camellia sinensis, and Trigonella foenum-graecum* are the plants that exhibit antioxidant action.

Some examples of antioxidants include flavonoids and tannins. Flavonoids are plant compounds with antioxidant properties which benefit health. Kaempferol is a natural flavanol, which can reduce the risk of cancer. It stimulates the body's antioxidants against free radicals that cause cancer. In addition to having antiviral and anti-allergic properties, flavonoids are also thought to protect against inflammation and tumor growth. Tannins, on the other hand, are type of plant compound that exerts high antioxidant activity and offers protection for the gastrointestinal system. The term "tannin" comes from its original use in tanning leathers. It is thought that the existence of tannins in some herbs can help relieve gastro- intestinal complaints like diarrhea. The properties of flavonoids and tannins can lower the risk of cardiovascular disease and cancer (Ullah et al. 2020).

3.5 Blood Purification

It is generally believed that if the blood is cleansed and excess acidity is neutralized, all ailments will eventually go away. As a result, blood purifiers play an important role in herbal remedies.

The body's blood and lymph transport a wide range of hazardous compounds, the majority of which are acids. These compounds include food and drink ingredients such as chemical preservatives that the body cannot simply reject. Natural wastes of the body are also included, which may be created in excess or insufficiently removed when the body's functions malfunction. Toxins in the blood are regarded excess "heat" in traditional Chinese medical philosophy, and toxin-producing infections are referred to as hot diseases. The small intestine, which must extract beneficial nutrients from the things taken, is the place of the body primarily responsible for blood purity. The liver, kidney, and colon are secondary organs that influence blood purification.

There are various methods for purifying the blood through herbal treatment:

1. Neutralizing acids immediately using the high alkalinizing effect of specific herbs (e.g., dandelion and slippery elm)
2. Boosting the body's key organic activities, particularly associated with liver, kidney, lungs, and intestines (e.g., Oregon grape root and goldenseal)
3. Eliminating extra fat and moisture where toxins are trapped (e.g., plantain, mullein, chickweed, and gotu kola)
4. Get rid of excess heat, particularly in small intestine (e.g., rhubarb root)

In addition to antioxidant and blood-purifying actions, herbs also possess anesthetic, antibacterial, antifungal, antiviral, anti-inflammatory, anticancer, sedative, and laxative properties. Other therapeutic properties of herbs have been summarized along with examples in Table 3.1.

Table 3.1 Therapeutic properties and healing actions of herbs

Properties	Action	Herbs	References
Anesthetic	An anesthetic or painkiller is any member of the group of drugs used to achieve analgesia, relief from pain, without losing consciousness	*Erythroxylum coca, Syzygium aromaticum, Cinchona officinalis, Valeriana officinalis, Aloe vera, Glycyrrhiza glabra,* and *Ocimum sanctum*	Vaishnavi et al. (2020)
Antibacterial	Inhibits or destroys the growth of bacteria	*Syzygium aromaticum, Taxus baccata, Terminalia arjuna, Tinospora cordifolia, Withania somnifera*	Dhama et al. (2014)
Antibiotic	Prevents the growth of microorganisms that cause infectious diseases	*Allium cepa, Allium sativum* (garlic) extracts, *Zingiber officinale, Ginkgo biloba, Syzygium aromaticum, Cinnamomum* spp., *Thymus vulgare,* and *Brassica campestris*	Sayed (2023)
Antifungal	Inhibits and destroys growth of fungi.	*Curcuma longa, Cymbopogon citratus, Allium sativum, Zingiber officinale, Ageratum conyzoides,* and *Nyctanthes arbor-tristis*	Nautiyal (2023)
Anti-inflammatory	Reduces or suppresses inflammation	*Curcuma longa, Zingiber officinale, Rosmarinus officinalis,* and *Borago officinalis*	Ghasemian et al. (2016)
Anticancer	Ease symptoms, reduce side effects, inhibit tumor growth and progression, and slower the spread of cancer	*Catharanthus roseus, Curcuma longa, Ferula asafoetida, Maclura pomifera, Papaver somniferum, Podophyllum peltatum, Raphanus sativus, Taxus brevifolia, Thymus vulgaris, Withania somnifera*	Greenwell and Rahman (2015)
Antiparasitic	Combats parasite activity	*Thymus vulgaris, Cuminum cyminum, Origanum vulgare,* and *Rosmarinus officinalis*	Strothmann et al. (2022)
Antiviral	Eliminates or inhibits the effects of viruses	*Phyllanthus amarus, Melissa officinalis, Sambucus nigra, Andrographis paniculata, Salvia officinalis,* and *Glycyrrhiza glabra*	Martin and Ernst (2003)
Decongestant	A treatment that relieves congestion	*Ginkgo biloba, Ephedra* sp., and *Glycyrrhiza glabra*	Lu and Lu (2014)
Diuretic	Increases the excretion of urine	*Mangifera indica, Lepidium sativum, Mimosa pudica,* and *Achyranthes aspera*	Dutta et al. (2014)
Immune Stimulation	Helps support the immune system in protecting against illness	*Phyllanthus emblica, Cinnamon verum, Withania somnifera,* and *Tinospora cordifolia*	Ara et al. (2020)
Sedative	Calms nervousness and anxiety	*Ziziphus spinosa, Glycyrrhiza uralensis,* and *Poria cocos*	Singh and Zhao (2017)
Laxative	Constipation reliever	*Prunus persica, Cyamopsis tetragonolobus, Citrus sinensis, Plantago ovata, Cassia auriculata, Euphorbia thymifolia,* and *Croton tiglium*	Akram et al. (2022)

3.6 Healing Properties of Some Common Herbs

3.6.1 Achyranthes aspera *L.*

Family: Amaranthaceae
   Common Name: Rough Chaff Flower
   Uses: The plant possesses antidiabetic and antirheumatic activities and also found to be effective against abdominal tumors. Seeds are beneficial for bleeding piles. Leaves are used for curing stomachache, skin ailments, boils, and abscess. Leaf paste is applied to cure insect bites. Extract of roots is used for dysentery and menstrual disorders.

3.6.2 Asparagus adscendens *Roxb.*

Family: Asparagaceae
   Common Name: Shatawari
   Uses: Root is diuretic, aphrodisiac, appetizer, astringent, and laxative. It is used for dysentery, diarrhea, leprosy, and throat infection. Leaves are used for headache, nausea, and vomiting. Root is used for vitality and strength.

3.6.3 Azadirachta indica *A. Juss.*

Family: Meliaceae
   Common Names: Margosa tree, neem tree
   Uses: Every part of the plant including seeds, leaves, and bark is used for medicinal purposes. It possesses antibiotic properties. Leaves are used to cure fever, skin diseases, and boils. Bark is used to cure fever and skin problems. Seed oil is used as an antiseptic and anthelmintic. Twigs are used as toothbrush which is found to be effective for bad breath, toothache, and gum problems.

3.6.4 Cannabis sativa *L.*

Family: Cannabaceae
   Common Names: Hemp, Marijuana
   Uses: It is antispasmodic, anesthetic, hypnotic, aphrodisiac, astringent, appetizer, sleep inducer, sedative, and narcotic. Leaves are used to cure inflammation, rheumatic pain, stomachache, diarrhea, wounds, and sores. Root powder is used to cure headache.

3.6.5 Cinnamomum camphora *(L.) Nees & Eberm*

Family: Lauraceae
  Common Name: Camphor tree
  Uses: The plant is a source of camphor, used for curing diarrhea, dysentery, wounds, ulcers, bronchitis, rheumatism, muscular pain, and pneumonia. It stimulates uterus, menstruation, and uterine hemorrhages.

3.6.6 Curcuma longa *L.*

Family: Zingiberaceae
  Common Name: Turmeric
  Uses: It is used for the treatment of liver disorders, dermatosis, piles, malignant ulcers, indigestion, edema, urinary complaints, bronchial asthma, anemia, leprosy, and chronic dysentery. It is a colorant, flavorant, and cosmetic. Turmeric is also used as blood purifier, stomachic, and tonic.

3.6.7 Glycyrrhiza glabra *L.*

Family: Fabaceae
  Common Names: Liquorice
  Uses: The herb is demulcent, expectorant, laxative, and sweetener. It is commonly used for the treatment of cough, hypertension, rheumatoid arthritis, dermatitis, and ulcers. As a sweetener, liquorice is used in chewing gums, chocolates, and baked food.

3.6.8 Ocimum sanctum *L.*

Family: Lamiaceae
  Common Name: Holy Basil
  Uses: Leaves are used for asthma, cough, fever, cold, bronchitis, skin eruptions, digestive disorders, catarrh, and dyspepsia. Decoction of roots is prescribed for malarial fever. Leaf juice is used for earache. Leaves possess diaphoretic, antiperiodic, stimulating, expectorant, antibacterial, and insecticidal properties.

3.6.9 Phyllanthus emblica *L.*

Family: Phyllanthaceae
  Common Name: Emblic Myrobalan
  Uses: It is diuretic, cardiac, astringent, laxative, refrigerant, and liver tonic. Fruit is rich source of vitamin C and used for the treatment of diarrhea, dysentery, constipation, diabetes, fever, headache, asthma, bronchitis, jaundice, cancer, inflammation of eyes, dyspepsia, leukorrhea, menorrhagia, indigestion, anemia, and hemorrhage.

3.6.10 Podophyllum hexandrum *Royle*

Family: Berberidaceae
  Common Name: Himalayan Mayapple
  Uses: The underground parts of the plant are stimulant, cholagogue, purgative, bitter tonic, hepatic, and alterative. Roots/rhizomes are used to cure ulcers, cuts, wounds, skin diseases, and cancer.

3.6.11 Ricinus communis *L.*

Family: Euphorbiaceae
  Common Name: Castor
  Uses: Castor oil is used as purgative. Roots are used diarrhea, constipated bowels, rheumatic affections, and gastrointestinal disorders. Seeds are used for promoting conception. Leaf juice is used for jaundice.

3.6.12 Taraxacum officinale *F.H. Wigg.*

Family: Asteraceae
  Common Name: Dandelion
  Uses: The well-known diuretic dandelion is a therapeutic herb. Being high in potassium, it maintains the body's essential potassium levels instead of lowering them as many prescribed diuretics do. As dandelion has a mild laxative effect and stimulates urine flow, it also acts as a good option for weight loss. Dandelion is rich in vitamin A and a strong electrolyte balancer due to its high potassium and organic

sodium content. When it comes to vitamin A content, dandelion greens have 7000 units per ounce, whereas lettuce and carrots have 1200 and 1275 units, respectively. In addition to supporting thyroid function and helping to increase white blood cell counts, vitamin A aids in normal cell reproduction in the body (Jalili et al. 2020).

3.6.13 *Thymus serpyllum L.*

Family: Lamiaceae
  Common Name: Wild Thyme
  Uses: It possesses *antiseptic, antispasmodic, anthelmintic, carminative, expectorant, stimulating, and diuretic* properties. It is commonly used for stomach disorders, nervous disorders, cough, and cold. Leaves and floral branches are used for epilepsy, whooping cough, and menstrual disorders. Leaves are used for the treatment of skin disorders.

3.6.14 *Tinospora cordifolia (Willd.) Hook.f. & Thomson*

Family: Menispermaceae
  Common Name: Guduchi
  Uses: It is anthelmintic, antiperiodic, anti-arthritic, aphrodisiac, antipyretic, blood purifier, cardiac, expectorant, digestive, carminative, and diuretic. It is used for curing jaundice, diarrhea, fever, cough, diabetes, and gout.

3.6.15 *Withania somnifera (L.) Dunal*

Family: Solanaceae
  Common Name: Ashwagandha
  Uses: Roots of the plant are diuretic, aphrodisiac, sedative, tonic, and abortifacient. They are used for the treatment of cough, rheumatism, stress, fatigue, arthritis, inflammation, ulcer and for promoting urination.

3.6.16 Zingiber officinale *Roscoe*

Family: Zingiberaceae
  Common Name: Ginger
  Uses: Since ancient times, *Zingiber officinale* has been used in Ayurvedic and traditional medicines to treat a wide range of conditions, such as common cold,

fever, sore throat, pain, rheumatism, bronchitis, digestive issues, gastrointestinal disorders, nausea, and vomiting. It is also used as a carminative and appetite stimulant.

3.7 Precautions for Herbal Treatment

Too much of anything, like with all prescriptions and over-the-counter treatments, is still too much. Herbs, like everything else, should be taken at recommended dosages, beginning with the lowest if never taken it before. Many individuals believe that herbs are safer than pharmaceuticals since they are natural; however, detractors of herbal treatments argue that pharmaceuticals are safer because the dose is more regulated and herb users must guess at the dosages they take.

The potency of a specific herb is determined by several factors, including the plant's genetics, the herb's growth environment, the plant's maturity, the technique of preparation, and the amount of time the herb has been stored. No one, however, can guarantee that prescription medications are taken appropriately.

Here are some guidelines for using therapeutic herbs (Ekor 2014):

1. Herbs, with the exception of a few, should not be given to children under the age of two, and even then, the infusion should be diluted.
2. Except for a few herbs, pregnant and breastfeeding women should avoid using therapeutic herbs. Even if the herb is safe to consume while pregnant, consultation with the specialist/ herbalist is required.
3. Special cautious are required if the patient is taking chronic drugs because herbs can interact with the prescriptions the patient is taking. An herbal cure, for example, may reduce the effectiveness of birth control tablets.
4. If any indicator of toxicity is noticed, the use of the plant should be discontinued immediately. Toxic symptoms include nausea, diarrhea, dizziness, and headache.

3.8 Herbal Products

Plant-derived items such as nuts, seeds, berries, hulls, shells, plant oils, and saps can be classified as herbal products. Nuts are nutrient-dense nuggets that are packed with fiber, protein, vitamin E, and plant sterols. Studies have shown that nuts protect against cholelithiasis, type 2 diabetes, obesity and retinal degeneration. Seeds are unique in that they have several trace minerals and phytochemicals as well as balanced fat ratios (Hever 2016). Barley, rice, wheat, and oats provide rich nutrients for the body to grow. High protein foods such as soybeans, almonds, sunflower seeds, and pumpkin seeds add the quality and provide the strength. Lecithin as well as vitamins A, B, C, F, and unsaturated fatty acids are found in seeds, nuts, and grains. Furthermore, auxin, naturally occurring chemical that delays premature aging through cell renewal is found in seeds, nuts, and cereals. Additionally, seeds contain pacifiers, which support the development of antibiotic responses and strengthen our innate defences against illnesses (Vincent 2011).

Honey is a food substance which is produced by honeybees from nectar and plant sweet deposits, and it helps to fight seasonal allergies. Honey can lessen the intensity of allergy symptoms and helps to develop immunity against seasonal allergies. It is traditionally used for the treatment of cough, asthma, wounds, burns, and cardiovascular and gastrointestinal disorders. In addition, honey helps with circulation issues, liver and kidney problems, and cold recovery.

The pollen that bees collect from flowers might be considered as herb due to its nutritional and medicinal properties. Bioactive substances included in bee pollen include lipids, proteins, amino acids, carbohydrates, minerals, vitamins, and polyphenols. The essential elements of bee pollen improve a variety of biological processes and provide immunity against numerous illnesses. Seasonal allergy symptoms are also lessened by bee pollen (Kalfa et al. 2009).

Chapter 4
Herbs of Bee Interest

4.1 Introduction

Honeybees are the social insect which are well known for excellent colony organization. During the evolutionary process, they have been evolved along with flowering plants (Angiosperms). Both are mutually benefitted as honeybees are dependent on flowering plants for their food requirements and in turn, helping in the pollination of plants. Honeybees are the prominent pollinator of the ecosystem which pollinate a wide variety of cultivated as well as wild plant species. They store pollen and nectar for future purposes. There are 4 main species of honeybee, viz. *Apis dorsata* (Rock bee), *Apis florea* (Little bee), *Apis cerana* (Asian bee), and *Apis mellifera* (European bee). In Indian perspective, *Apis mellifera* and *Apis cerana* are usually practiced for commercial beekeeping. *Apis mellifera* which is also known as Italian honeybee is generally preferred by the beekeepers because it gives comparatively more yield and is found to be less aggressive than the other species.

Beekeeping plays a significant role in food security and conservation as well as management of natural resources through pollination. It helps in creating better opportunities for employment, contributing in social and economic development. The major factor responsible for successful beekeeping is the health of colony which depends upon the availability of food. Honeybees are dependent on local flora for their food requirements. Among 2,50,000 plant species in the world, approximately 40,000 species are the important source of food for honeybees. The diversity of bee flora includes forest trees, grasses, shrubs, climbers as well as cultivated plants which contribute significantly to beekeeping. The flowering plants include both nectariferous and polleniferous plants providing food to honeybees in the form of nectar and pollen, respectively. Natural resin from the plants called propolis is also collected by the hive bees and the sting less bees for building their nest and for defense as well. The wide variety of plant species from which honeybees collect pollens and nectar constitute bee flora, also called bee forage or bee

Fig. 4.1 Essential qualities of bee flora

pasture. Nectar is the source of carbohydrates, and pollens are the source of proteins for honeybees. Nectar is secreted in the flowers by special glands called nectaries. It is generally used by the plants to attract insects for pollination. Factors affecting quality and production of nectar are sunlight, wind, temperature, relative humidity, soil conditions, etc. Pollens are produced inside the anther of flower. They are generally used by the honeybees for rearing of their brood. Pollen and nectar availability to foraging bees varied with time of the season and flowering periods of different plants species. The flora of a region depends on its agroclimatic conditions and generally varies from place to place. Therefore, for enhancing the production of honey, it is essential to have knowledge pertaining to the length of flowering period, seasonal availability, and flowering phenology of bee flora (Fig. 4.1).

4.2 Floral Calendar

For harvesting maximum honey, beekeepers have to carefully observe and identify the prominent nectar producing and pollen yielding plants by recording their flowering periods and their relevance to the development of colony. Floral calendar is a kind of timetable which indicates the time and duration of blossom period of plants yielding nectar and pollens in a region. It represents the seasonal variation in bee flora. It is an important tool for the efficient management of bee keeping so that beekeepers can get maximum benefit. Success of the bee keeping industry depends on the availability, quantity, and quality of raw material (nectar and pollen of the flower). The floral calendar includes the scientific and common names of the plant species, flowering period, source of pollen, and nectar for honeybees as well as its density and distribution in the region. It helps in seasonal management practices for bee keeping and also to identify the floral peaks and dearth. It ultimately helps to improve the bee forage resources in the region which enables the establishment of an appropriate number of colonies in that area.

4.3 Bee Flora

Honeybees are dependent on flowering plants for their nutritional needs. The diverse flora of bee interest includes **Medicinal plants** (*Aegle marmelos, Azadirachta indica, Bergera koenigii, Butea monosperma, Echinops echinatus, Justicia adhatoda, Mesua ferrea, Ocimum tenuiflorum, Ricinus communis, Tamarindus indica, Tylophora asthmatica*); **Ornamental plants** (*Antigonon leptopus, Hamelia patens, Helianthus annuus, Nerium indicum, Pongamia pinnata, Tagetes erecta*); **Horticultural plants** (*Annona squamosa, Carica papaya, Citrullus lanatus, Citrus limon, Cocos nucifera, Emblica officinalis, Mangifera indica, Manilkara zapota, Phoenix dactylifera, Psidium guajava, Syzygium jambos, Tamarindus indica, Ziziphus mauritiana*); **Vegetables** (*Abelmoschus esculentus, Allium cepa, Amaranthus gracilis, Cucumis sativus, Cucurbita pepo, Lagenaria siceraria, Luffa acutangular, Lycopersicon esculentum, Medicago sativa, Momordica charantia, Moringa oleifera, Phaseolus vulgaris, Pisum sativum, Solanum melongena, Trichosanthes anguina*); **Pulses**(*Cajanus cajan, Cicer arietinum, Phaseolus vulgaris, Vigna mungo, Vigna radiata*); **Plantation/Trees** (*Acacia* spp., *Albizia lebbeck, Casia fistula, Dalbergia sissoo, Delonix regia, Eucalyptus* spp., *Toona ciliata*); **Weeds** (*Ageratum conyzoides, Argemone mexicana, Bidens pilosa, Cosmos* spp., *Lantana camara, Leucas aspera, Parthenium hysterophorus, Trianthema portulacastrum*) and other **field crops** (*Arachis hypogaea, Brassica rapa, Coffea arabica, Coriandrum sativum, Eruca sativa, Gossypium* spp., *Hordeum vulgare, Oryza sativa, Pennisetum typhoides, Sesamum indicum, Trifolium alexandrinum, and Zea mays*) (Figs. 4.2 and 4.3).

50   4 Herbs of Bee Interest

Fig. 4.2 Some common honey plant resources

Fig. 4.3 Some common honey plant resources (cond...)

4.4 Description of Some Common Honey Plant Resources

Abelmoschus esculentus **(L.) Moench**
Syn: *Abelmoschus longifolius* (Willd.) Kostel.; *A. praecox* (Forssk.) Sickenb.; *Hibiscus esculentus* L.; *H. longifolius* Roxb.
Family: Malvaceae
Common Name: Okra
Distribution: It is native to India and Myanmar and widely cultivated in tropical and subtropical regions.
Flowering & Fruiting Season: April–September
Source of Bee Forage: N (Nectar), P (Pollen)
Description: It is an erect annual herb. Leaves are alternate, dark green, hairy, 3–7 lobed. Solitary flowers arise in the axils of leaves. Fruit is an oblong to oblong-ovate, 5-angled capsule with dark brown, glabrous seeds.

Ageratum conyzoides **L.**
Syn: *Ageratum album* Steud.; *A. ciliare* L.; *A. cordifolium* Roxb.; *A. hirsutum* Poir; *Eupatorium conyzoides* (L.) E.H.L. Krause
Family: Asteraceae
Common Name: *Billygoat Weed*
Distribution: It is native to Mexico. It is naturalized in Asia, Australia, and Africa as invasive weed.
Flowering & Fruiting Season: December–May
Source of Bee Forage: P
Description: It is an erect annual herb. Leaves opposite, ovate having serrate margins. Flowers are arranged in terminal heads. Fruit is small, single seeded.

Albizia lebbeck **(L.) Benth.**
Syn: *Albizia latifolia* Boivin; *A. speciosa* (Jacq.) Benth.; *Acacia lebbeck* (L.) Willd.; *Mimosa lebbeck* L.
Family: Fabaceae
Common Name: Indian Siris
Distribution: It is indigenous to India, Bangladesh, Myanmar, Nepal, Sri Lanka, and Pakistan.
Flowering & Fruiting Season: March–May
Source of Bee Forage: NP
Description: It is a medium-sized, deciduous tree with spreading crown. Leaves are alternate, bipinnate, and stipulate. Flowers are bisexual, arranged in subglobose heads. Fruit is an oblong, flat pod.

Allium cepa **L.**
Syn: *Allium angolense* Baker; *A. aobanum* Araki; *A. ascalonicum* Anon.; *Cepa esculenta* Gray; *C. vulgaris* Garsault; *Kepa esculenta* Raf.
Family: Amaryllidaceae
Common Name: Onion
Distribution: Cultivated
Flowering & Fruiting Season: May–June

Source of Bee Forage: NP

Description: It is a perennial herb with underground bulb. Radical leaves are cylindrical and fleshy. Flowers are small, white to pink in color, arranged in umbels. Fruit is a loculicidal capsule.

Allium sativum **L.**

Syn: *Allium arenarium* Sadler ex Rchb.; *A. longicuspis* Regel; *A. ophioscorodon* Link; *A. pekinense* Prokh.; *Porrum sativum* (L.) Rchb.

Family: Amaryllidaceae

Common Name: Garlic

Distribution: It is native to Iran, Turkmenistan, Kazakhstan, Kirgizstan, Uzbekistan, and Tadzhikistan.

Flowering & Fruiting Season: May–July

Source of Bee Forage: NP

Description: It is a perennial herb with underground bulb. Leaves are long, flattened, sword-shaped. Flowers are greenish-white or pink in color arising in dense clusters.

Althaea officinalis **L.**

Syn: *Althaea balearica* J.J. Rodr.; *A. multiflora* Rchb. ex Regel; *A. sublobata* Stokes; *Malva althaea* E.H.L. Krause; *M. maritima* Salisb.

Family: Malvaceae

Common Name: Hollyhock

Distribution: It is native to Europe, Northern Africa, and Western Asia. It is also grown as ornamental plant.

Flowering & Fruiting Season: April–July

Source of Bee Forage: NP

Description: It is a branched perennial herb. Leaves are simple, ovate, or heart-shaped, alternate with toothed margins. Flowers are pink to red in color, arising singly or in short, dense axillary clusters.

Argemone mexicana **L.**

Syn: *Argemone alba* Raf.; *A. leiocarpa* Greene; *A. mexicana* var. *typica* Prain; *Papaver mexicanum* (L.) E.H.L. Krause

Family: Papaveraceae

Common Name: Mexican Prickly Poppy

Distribution: It is indigenous to tropical America. It is naturalized as an agricultural weed throughout tropical and subtropical regions.

Flowering & Fruiting Season: May–July

Source of Bee Forage: NP

Description: It is an annual glabrous herb having prickly branched stem. Leaves are oblong, alternate, serrate with spiny margins. Flowers are solitary, terminal, yellow in color. Fruit is a prickly capsule.

Azadirachta indica **A. Juss.**

Syn: *Azadirachta indica* var. *minor* Valeton; *Melia azadirachta* L.; *M. indica* (A. Juss.) Brandis; *M. japonica* Hassk.; *M. pinnata* Stokes

Family: Meliaceae

Common Name: Neem

4.4 Description of Some Common Honey Plant Resources

Distribution: It is native to Bangladesh, Assam, Myanmar, Cambodia, Thailand, and Vietnam.
Flowering & Fruiting Season: April–May
Source of Bee Forage: NP
Description: It is an evergreen tree. Leaves are alternate, pinnate, and petiolate. Inflorescence is an axillary, many-flowered thyrsus. Flowers are small, white in color, bisexual or staminate, arising in loose clusters. Fruit is a drupe.

Bergera koenigii L.
Syn: *Camunium koenigii* (L.) Kuntze; *Chalcas koenigii* (L.) Kurz; *Murraya koenigii* (L.) Spreng.; *M. siamensis* Craib
Family: Rutaceae
Common Name: Curry leaf tree
Distribution: It is native to Asia, distributed in tropical and subtropical regions.
Flowering & Fruiting Season: April–May
Source of Bee Forage: P
Description: It is a shrub or small tree. Leaves are imparipinnate having 11–25, alternate, acuminate leaflets. Flowers are white arranged in terminal corymbose cymes. Fruits are ovoid or subglobose, 2-seeded.

Brassica campestris L.
Syn: *Brassica rapa* L.; *B. campestris* var. *oleifera* de Candolle; *B. pekinensis* (Loureiro) Ruprecht; *B. chinensis* L.; *B. rapa* subsp. *chinensis* (L.) Hanelt; *Sinapis pekinensis* Loureiro
Family: Brassicaceae
Common Name: Mustard
Distribution: It is widely cultivated as an oil crop and vegetable.
Flowering & Fruiting Season: January–March
Source of Bee Forage: NP
Description: An erect herb. Lower leaves are short-petiolate, obovate, and upper leaves are oblong, sessile, obtuse-rounded at the apex. Flowers with four yellow petals are arranged in racemes. Seeds are globose, dark brown in color.

Cajanus cajan (L.) Huth
Syn: *Cajanus flavus* DC.; *C. bicolor* DC.; *C. indicus* Spreng.; *C. luteus* Bello; *Cytisus cajan* L.; *C. pseudocajan* Jacq.
Family: Fabaceae
Common Name: Pigeon pea
Distribution: It is widely cultivated in tropical and subtropical regions.
Flowering & Fruiting Season: April–May
Source of Bee Forage: NP
Description: It is an erect, perennial shrub. Leaves are trifoliate having elliptic-lanceolate leaflets. Yellow colored flowers arise in racemes. The fruit is a pod with round or lens-shaped seeds.

Campsis radicans (L.) Bureau.
Syn: *Bignonia radicans* L.; *B. coccinea* Steud.; *B. florida* Salisb.; *Tecoma radicans* (L.) Duhamel; *T. speciosa* Parsons
Family: Bignoniaceae

Common Name: Trumpet Creeper
Distribution: It is native to southeastern United States and naturalized in several parts of the world.
Flowering & Fruiting Season: June–July
Source of Bee Forage: NP
Description: It is a deciduous woody vine. Leaves are opposite, compound, elliptic-oblong with serrate margins. Flowers are orange-colored, trumpet-shaped, arranged in panicles. Fruit is a many-seeded capsule.

Cannabis sativa L.
Syn: *Cannabis americana* Pharm. ex Wehmer; *C. chinensis* Delile; *C. indica* Lam.; *C. macrosperma* Stokes; *C. sativa* var. *vulgaris* Alef.
Family: Cannabaceae
Common Name: Bhang
Distribution: It is native to Afghanistan, Pakistan, Xinjiang, Kazakhstan, Tadzhikistan, Kirgizstan, Uzbekistan, and Turkmenistan.
Flowering & Fruiting Season: June–October
Source of Bee Forage: P
Description: It is an annual herb with branched stem. Leaves are stipulate, palmately compound having serrate, lanceolate leaflets. Flowers are unisexual. Fruit is a small achene.

Carica papaya L.
Syn: *Carica citriformis* J. Jacq. ex Spreng.; *C. peltata* Hook. & Arn.; *C. posopora* L.; *C. sativa* Tussac; *Papaya carica* Gaertn.
Family: Caricaceae
Common Name: Papaya
Distribution: It is native to tropical America, cultivated in tropical and subtropical region of the world.
Flowering & Fruiting Season: July–August
Source of Bee Forage: NP
Description: It is a soft-wooded tree crowned with palmately lobed leaves having hollow petiole. In the axils of the leaves, three types of flowers borne in modified cymose inflorescence: staminate, pistillate, and hermaphrodite. Succulent fruits have numerous black colored seeds.

Cassia fistula L.
Syn: *Cassia rhombifolia* Roxb.; *C. excelsa* Kunth; *C. rhombifolia* Roxb.; *Cathartocarpus rhombifolius* (Roxb.) G.Don
Family: Fabaceae
Common Name: Amaltas
Distribution: It is commonly found in tropical and subtropical regions, up to 1200 m.
Flowering & Fruiting Season: May–June
Source of Bee Forage: NP
Description: It is a small tree. Leaves are pinnately compound having 4–8 pairs of leaflets. Flowers are bright yellow in color arranged in hanging racemes. Fruits are cylindrical with compressed seeds.

4.4 Description of Some Common Honey Plant Resources

Cicer arietinum L.
Syn: *Vicia arietina* (L.) E.H.L. Krause
Family: Fabaceae
Common Name: Chickpea, Bengal Gram
Distribution: It is native to Iran, Iraq, and Turkey. It is cultivated in many parts of the world.
Flowering & Fruiting Season: February- March
Source of Bee Forage: NP
Description: It is an annual, erect, branched herb having pinnately compound leaves. Leaflets are ovate oblong or obovate. Flowers are axillary and solitary. Fruit is an oblong pubescent pod.

Citrullus lanatus (Thunb.) Matsum. & Nakai
Syn: *Citrullus vulgaris* var. *lanatus* (Thunb.) L.H. Bailey; *C. aquosus* Schur; *C. battich* Forssk.; *Momordica lanata* Thunb.
Family: Cucurbitaceae
Common Name: Watermelon
Distribution: It is native to Southwestern Africa.
Flowering & Fruiting Season: April–June
Source of Bee Forage: NP
Description: It is a prostrate, annual, hairy vine with pinnately lobed leaves and branched tendrils. Flowers are axillary, pale-green in color. Fruit is glabrous, spherical to oval in shape.

Crotalaria juncea L.
Syn: *Crotalaria benghalensis* Lam.; *C. cannabinua* Royle; *C. fenestrate* Sims; *C. sericea* Willd.; *C. tenuifolia* Roxb. ex Hornem.
Family: Fabaceae
Common Name: Sun Hemp
Distribution: It is a tropical and subtropical legume cultivated in many countries.
Flowering & Fruiting Season: September–October
Source of Bee Forage: N
Description: It is an erect branched herb with cylindrical, ribbed stem. Leaves are simple, hairy, elliptic, or oblong-lanceolate. Flowers are yellow in color arranged in terminal raceme. Fruit is a cylindrical, many-seeded pod.

Cucumis melo L.
Syn: *Cucumis deliciosus* Salisb.; *C. alba* Nakai; *C. acidus* Jacq.; *C. aromaticus* Royle; *Bryonia collosa* Rottler
Family: Cucurbitaceae
Common Name: Muskmelon
Distribution: It generally grows in dry tropical regions.
Flowering & Fruiting Season: April–June
Source of Bee Forage: NP
Description: It is a branched, annual, hairy vine with simple, alternate leaves, and unbranched tendrils. The axillary flowers may be gynoecious (female), monoecious (male and female), or perfect. Fruit is round to ellipsoid, fleshy berry.

Cucumis sativus L.

Syn: *Cucumis esculentus* Salisb.; *C. hardwickii* Royle; *C. muricatus* Willd.; *C. rumphii* Hassk.; *C. vilmorinii* Spreng.
Family: Cucurbitaceae
Common Name: Cucumber
Distribution: It is native to Bangladesh, China, Myanmar, Nepal, Thailand, and Eastern and Western Himalayas. It is commonly cultivated throughout the world.
Flowering & Fruiting Season: May–September
Source of Bee Forage: NP
Description: It is an annual vine with simple, alternate leaves, and unbranched tendrils. Axillary male and female flowers arise separately on the same plant. Fruit is many-seeded, cylindrical berry.

Cucurbita pepo **L.**
Syn: *Cucurbita aurantia* Willd.; *C. courgero* Ser.; *C. elongata* Bean ex Schrad.; *C. esculenta* Gray; *C. melopepo* L.; *C. subverrucosa* Willd.; *C. verrucosa* L.; *Pepo citrullus* Sag.
Family: Cucurbitaceae
Common Name: Marrow, Pumpkin
Distribution: It is native to Eastern United States and Mexico and cultivated worldwide.
Flowering & Fruiting Season: April–June
Source of Bee Forage: NP
Description: It is a creeping herb having simple, alternate, palmately lobed leaves with branched tendrils. Unisexual flowers are yellow in color. Fruits are berries having variable shapes.

Eriobotrya japonica **(Thunb.) Lindl.**
Syn: *Crataegus bibas* Lour.; *Mespilus japonica* Banks.; *Rhaphiolepis loquata* B.B. Liu & J. Wen
Family: Rosaceae
Common Names: Loquat, The Chinese Medlar
Distribution: It is native to China. Generally cultivated elsewhere.
Flowering & Fruiting Season: January - February
Source of Bee Forage: NP
Description: It is a small evergreen tree. Leaves are leathery, alternate, elliptic-oblong. Flowers are white in color, arranged in panicles. Fruits are oval or rounded, arising in clusters, each containing 3–5 large seeds.

Eruca sativa **Mill.**
Syn: *Eruca eruca* (L.) Asch. & Graebn.; *E. oleracea* J.St.-Hil.; *E. vesicaria* subsp. *sativa* (Miller) Thell.; *Brassica eruca* L.; *B. hispida* Ten.; *Sinapis eruca* (L.) Vest
Family: Brassicaceae
Common Name: Taramira
Distribution: It is native to Mediterranean region, China, and Arabian Peninsula.
Flowering & Fruiting Season: February–March
Source of Bee Forage: NP

4.4 Description of Some Common Honey Plant Resources

Description: It is an erect, branched herb with slender taproot. Leaves are pinnately lobed. Upper leaves are petiolate, toothed. Flowers are arranged in racemes. Fruit is a siliqua. Seeds are flattened, brownish in color.

Eucalyptus tereticornis Sm.

Syn: *Eucalyptus insignis* Naudin; *E. populifolia* Desf.; *E. tereticornis* var. *pruiniflora* (Blakely) Cameron
Family: Myrtaceae
Common Name: Forest Red Gum
Distribution: It is native to Australia and New Guinea.
Flowering & Fruiting Season: September–December
Source of Bee Forage: NP
Description: It is a branched, tall tree. Leaves are simple, alternate, ovate, elliptic, lanceolate, with entire margins. Flowers are bisexual, white in color, arising in axillary umbels. Fruit is a many-seeded capsule.

Foeniculum vulgare Mill.

Syn: *Foeniculum capillaceum* Gilib.; *F. divaricatum* Griseb.; *F. dulce* Mill.; *Anethum dulce* DC.; *A. foeniculum* L.
Family: Apiaceae
Common Name: Fennel
Distribution: It is native to southern Europe and the Mediterranean region.
Flowering & Fruiting Season: April–May
Source of Bee Forage: NP
Description: It is a perennial herb having soft, feathery foliage. Small yellow-colored flowers arise in umbels. Fruit is oblong-ovoid, glabrous.

Hamelia patens Jacq.

Syn: *Hamelia coccinea* Sw.; *H. erecta* Jacq.; *Duhamelia odorata* Willd. ex Schult.; *Lonicera verticillata* (Mill.) Steud.
Family: Rubiaceae
Common Name: Hummingbird Bush, Scarlet Bush
Distribution: It is native to tropical America. It is commonly cultivated as ornamental plant.
Flowering & Fruiting Season: July–November
Source of Bee Forage: P
Description: It is an evergreen bushy plant. Leaves are elliptic, ovate, whorled. Flowers are tubular, bright red or scarlet in color. Fruit is a several seeded berry.

Helianthus annuus L.

Syn.: *Helianthus annuus* var. *apicalis* Cockerell; *H. aridus* Rydb.; *H. macrocarpus* DC.; *H. ovatus* Lehm.; *H. multiflorus* L.
Family: Asteraceae
Common Name: Sunflower
Distribution: *Helianthus annuus* is native to North America. It is widely cultivated throughout the world.
Flowering & Fruiting Season: May–September
Source of Bee Forage: NP

Description: It is a tall, hairy, annual herb. Leaves are simple, petiolate, alternate with serrate margins. The flowerhead has many small flowers. The outer ray flowers are yellow in color, and the central disk flowers are yellow to brown in color. The fruit is a single-seeded achene.

***Iberis amara* L.**

Syn: *Iberis affinis* Jord.; *Biauricula amara* (L.) Bubani; *B. panduriformis* Bubani; *Thlaspi amarum* (L.) Crantz

Family: Brassicaceae

Common Name: Candytuft

Distribution: It is native to France, Britain, Germany, Italy, Belgium, Spain, and Switzerland.

Flowering & Fruiting Season: January–March

Source of Bee Forage: NP

Description: It is a branched erect annual herb. Leaves alternate, petiolate, or sessile. Bisexual flowers are arranged in corymbs. Fruit is an ovate-suborbicular silicula with flattened, winged seeds.

***Isodon rugosus* (Wall. ex Benth.) Codd**

Syn: *Plectranthus rugosus* Wall. ex Benth.; *Rabdosia rugosa* (Wall. ex Benth.) H. Hara; *Isodon plectranthoides* Schrad. ex Benth.; *Ocimum densiflorum* Roth.

Family: Lamiaceae

Common Names: Plectranthus, Wrinkled Leaf Isodon

Distribution: It is native to Oman, Bangladesh, India, Pakistan, Nepal, Afghanistan, Eastern and Western Himalaya, generally found in subtropical regions.

Flowering & Fruiting Season: July–September

Source of Bee Forage: N

Description: It is an erect branched shrub having opposite, tomentose, rugose, petiolate leaves with obtuse apex. Cymes are axillary arranged on lateral branches. Nutlets are oblong, brownish in color.

***Justicia adhatoda* L.**

Syn.: *Justicia caracasana* Sieber ex Nees; *Adhatoda arborea* Raf.; *A. pubescens* Moench; *A. vasica* Nees; *A. zeylanica* Medik.

Family: Acanthaceae

Common Name: Malabar nut, Vasaka

Distribution: It is distributed in tropical and subtropical regions up to 1300 m.

Flowering & Fruiting Season: February–April

Source of Bee Forage: NP

Description: It is a wild evergreen shrub. Leaves are ovate, lanceolate, acute, or acuminate at apex. Flowers are arranged in axillary spikes. Fruit is a subacute 4-seeded capsule.

***Lantana camara* L.**

Syn: *Lantana camara* subsp. *aculeata* (L.) R.W. Sanders; *L. camara* subsp. *glandulosissima* (Hayek) R.W. Sanders; *Camara vulgaris* Benth.

Family: Verbenaceae

Common Name: Lantana

Distribution: It is native to tropical America. It is cultivated as ornamental plant.

Flowering & Fruiting Season: March–September
Source of Bee Forage: NP
Description: It is an erect spiny shrub with angular stem and branches. Leaves are simple, petiolate, opposite decussate. Flowers are complete, hermaphrodite arranged in umbels or compound spikes. Fruit is a drupe.

Mangifera indica **L.**
Syn: *Mangifera domestica* Gaertn.; *M. gladiate* Bojer; *M. indica* var. *domestica* (Gaertn.) Blume; *M. racemosa* Bojer
Family: Anacardiaceae
Common Name: Mango
Distribution: It is native to India (Assam) China, Eastern Himalaya, Myanmar, and Thailand. It is generally cultivated in tropical and warmer subtropical regions.
Flowering & Fruiting Season: March–April
Source of Bee Forage: NP
Description: It is a large evergreen tree having umbrella-shaped crown. Leaves are simple, alternate, petiolate, oblong-lanceolate. Flowers are small, greenish-white, arranged in branched panicles. Fruit is a fleshy drupe.

Medicago sativa **L.**
Syn: *Medicago sativa* subsp. *vulgaris* (Alef) Arcang.; *M. sativa* var. *vulgaris* Alef; *M. vera* Kirschl.
Family: Fabaceae
Common Name: Alfalfa, Lucerne
Distribution: It is native to *Africa, Asia and Europe*. It is widely cultivated and also naturalized in different parts of the world.
Flowering & Fruiting Season: March–May
Source of Bee Forage: NP
Description: It is an erect perennial herb. Leaves trifoliate, stipulate, coarsely toothed. Flowers are arranged in dense racemes. Fruit is a curled pod having 2–6 seeds.

Melaleuca citrina **(Curtis) Dum.Cours.**
Syn: *Callistemon citrinus* (Curtis) Skeels; *C. cunninghamii* K.Koch; *C. lanceolatus* Sweet; *Metrosideros citrina* Curtis; *M. lanceolata* Sm.
Family: Myrtaceae
Common Name: Bottle brush
Distribution: It is native to Australia and cultivated throughout the world as ornamental plant.
Flowering & Fruiting Season: March–April
Source of Bee Forage: N
Description: It is a shrub or small tree having lanceolate leaves. Flowers are crimson red in color arranged in dense spikes. Fruit is a globose capsule with numerous seeds.

Morus alba **L.**
Syn: *Morus alba* var. *arabica* Bureau; *M. alba* var. *elongata* Risso; *M. alba* var. *integrifolia* K.Koch; *M. dulcis* Royle
Family: Moraceae

Common Name: Mulberry
Distribution: It is native to China
Flowering & Fruiting Season: February- March
Source of Bee Forage: P
Description: It is a deciduous tree having a spreading crown. Leaves are alternate, ovate, simple to 3-lobed, dentate. Staminate and pistillate flowers are arranged in catkins. Fruits are black, white, or purple colored drupes.

Ocimum tenuiflorum **L.**
Syn: *Ocimum flexuosum* Blanco; *O. hirsutum* Benth.; *O. sanctum* L; *O. tomentosum* Lam.; *Plectranthus monachorum* (L.) Spreng.
Family: Lamiaceae
Common Name: Holy Basil
Distribution: It is native to tropical and subtropical Asia.
Flowering & Fruiting Season: August–September
Source of Bee Forage: NP
Description: It is an erect, branched, aromatic herb with 4-angled stem. Leaves are simple, opposite oblanceolate with serrate margins. Flowers are small, tubular, bisexual, arising in terminal spikes.

Pennisetum typhoides **(Burm. f.) Stapf & C.E. Hubb.**
Syn: *Pennisetum american* (L.) Leeke; *P. glaucum* (L.) R.Br.; *Cenchrus spicatus* (L.) Cav.; *C. americanus* (L.) Morrone
Family: Poaceae
Common Name: Bajra, Pearl millet
Distribution: Widely cultivated and generally naturalized along roadside.
Flowering & Fruiting Season: June–July
Source of Bee Forage: P
Description: It is an annual herb. Leaves are long, linear-lanceolate. Tubular flowers are arranged in dense racemes. Fruits are small, rounded grains, generally white, yellow, or brown in color.

Pisum sativum **L.**
Syn: *Pisum arvense* L.; *P. humile* Boiss. & Noe; *P. sativum* L. ssp. *arvense* (L.) Poiret; *P. sativum* L. var. *macrocarpon* Ser.
Family: Fabaceae
Common Name: Pea
Distribution: Cultivated.
Flowering & Fruiting Season: January–March
Source of Bee Forage: NP
Description: It is an erect climbing herb. Leaves are alternate, pinnately compound, petiolate. Complete, bisexual flowers are arranged in axillary racemes. Fruit is a legume.

Pongamia pinnata **(L.) Pierre**
Syn: *Pongamia glabra* Vent.; *P. pinnata* var. *minor* (Benth.) Domin; *P. pinnata* var. *typica* Domin; *Millettia pinnata* (L.) Panigrahi; *Cytisus pinnatus* L.
Family: Fabaceae
Common Name: Pongam Tree

4.4 Description of Some Common Honey Plant Resources

Distribution: It is cultivated as ornamental tree, native to tropical and subtropical Asia.
Flowering & Fruiting Season: March- April
Source of Bee Forage: N
Description: It is a medium-sized tree. Leaves are alternate, compound, oblong, imparipinnate. Flowers are bisexual arranged in axillary racemes. Fruit is a flattened pod having single seed.

Prunus cerasoides **Buch.-Ham. ex D.Don**
Syn: *Prunus carmesina* H.Hara; *P. puddum* (Roxb. ex Ser.) Miq.; *Cerasus pectinata* Spach; *C. phoshia* Buch.-Ham. ex D.Don; *C. puddum* Roxb. ex Ser.; *Maddenia pedicellata* Hook.f.
Family: Rosaceae
Common Name: Himalayan Cherry, Indian Wild Cherry
Distribution: It is native to Eastern and Western Himalaya, India, Nepal, Thailand, and Myanmar.
Flowering & Fruiting Season: October–November
Source of Bee Forage: NP
Description: It is a medium-sized tree. Leaves are glabrous, elliptic-ovate, acuminate. Flowers are bisexual, arising in fascicles. Fruit is an ellipsoid, one-seeded drupe.

Psidium guajava **L.**
Syn: *Psidium pyriferum* L.; *P. angustifolium* Lam.; *Guajava pumila* (Vahl) Kuntze; *G. pyrifera* Kuntze; *Myrtus guajava* (L.) Kuntze
Family: Myrtaceae
Common Name: Guava
Distribution: It is cultivated throughout the tropical regions.
Flowering & Fruiting Season: May–June
Source of Bee Forage: NP
Description: It is an evergreen small tree. Leaves are simple, opposite, elliptic to oblong. Flowers are white, solitary or in clusters. The fruit is a fleshy, ovoid berry.

Pyrus pashia **Buch.-Ham. ex D.Don**
Syn: *Pyrus pashia var. grandiflora* Cardot; *P. pashia var. obtusata* Cardot; *Malus pashia* (Buch.-Ham. ex D.Don) *Wenz.*
Family: Rosaceae
Common Name: Indian Wild Pear
Distribution: It is found in the Himalayan region at an altitude of 700–2700 m.
Flowering & Fruiting Season: March–April
Source of Bee Forage: NP
Description: It is a medium-sized deciduous tree. Leaves are ovate, lance-shaped, finely toothed. Flowers are white, fragrant, arranged in corymbs. Fruit is round, dark brown in color.

Raphanus sativus **L.**
Syn: *Raphanus acanthiformis* Morel ex Sasaki; *R. candidus* Vorosch.; *R. gayanus* (Fisch. & C.A. Mey.) G. Don ex Sweet
Family: Brassicaceae

Common Name: Radish
Distribution: Cultivated.
Flowering & Fruiting Season: April–May
Source of Bee Forage: NP
Description: It is an annual herb with erect leafy stem. Basal leaves are oblanceolate to oblong, uppermost leaves are simple, sublinear. Flowers are white, bisexual arranged in raceme corymbs. Fruits are erect cylindrical pods with ovoid seeds.

Ricinus communis **L.**
Syn: *Ricinus africanus* Mill.; *inermis* Jacq.; *R. speciosus* Burm.; *R. spectabilis* Blume.; *R. virdis* Willd.; *Croton spinosus* Linn.
Family: Euphorbiaceae
Common Name: Castor oil plant
Distribution: It is indigenous to tropical Africa.
Flowering & Fruiting Season: November–March
Source of Bee Forage: P
Description: It is an evergreen shrub. Leaves are large, simple, alternate, with toothed margins. Flowers are arranged in terminal spikes. Fruit is 3-lobed, spiny capsule.

Sesamum indicum **L.**
Syn: *Sesamum mulayanum* N. C. Nair; *S. orientale* L.; *S. luteum* Retz.; *S. malabaricum* Burm.; *S. occidentalis* Heer & Regel
Family: Pedaliaceae
Common Name: Til
Distribution: It is widely cultivated in tropical and subtropical regions of the world. It is native to Indian sub-continent.
Flowering & Fruiting Season: July–August
Source of Bee Forage: NP
Description: It is an erect annual herb. Leaves are ovate, hairy. Flowers are bell-shaped, solitary axillary. The fruit is a four-angled capsule. Seeds are small, flat, black or white in color.

Sorghum vulgare **Pers.**
Syn: *Sorghum bicolor* (L.) Moench; *Andropogon bicolor* (L.) Roxb.; *A. sorghum* (L.) Brot.; *Holcus bicolor* L.; *H. sorghum* L.
Family: Poaceae
Common Name: Sorghum
Distribution: It is originated in central Africa and cultivated in Europe, Australia, India, China and Pakistan.
Flowering & Fruiting Season: July–August
Source of Bee Forage: P
Description: It is a tall, erect grass. Leaves are alternate, long, narrow. Flowers are small, arranged in panicles. Grains are small, round, arising in clusters at the top of stem.

***Syzygium cumini* (L.) Skeels**

4.4 Description of Some Common Honey Plant Resources

Syn: *Eugenia jambolana* Lam.; *Myrtus cumini* L.; *Syzygium jambolanum* (Lam.) DC.; *Calyptranthes jambolana* (Lam.) Willd.; *Eugenia cumini* (Linn.) Druce.; *Eugenia caryophyllifolia* Lam.

Family: Myrtaceae

Common Name: Jamun

Distribution: It is native to India, Africa, China, Indonesia, Malaysia, generally found in tropical and subtropical regions.

Flowering & Fruiting Season: May–June

Source of Bee Forage: NP

Description: It is a large evergreen tree. Leaves are opposite, oblong-ovate to elliptic. Flowers are greenish-white in color, arranged in axillary or terminal panicles. Fruits are oblong, fleshy, single-seeded berries.

Tagetes erecta L.

Syn: *Tagetes major* Gaertn.; *T. corymbosa* Sweet; *T. elongata* Willd.; *T. patula* L.; *T. tenuifolia* Kunth.

Family: Asteraceae

Common Name: Marigold

Distribution: It is commonly cultivated as garden plant.

Flowering & Fruiting Season: January–March

Source of Bee Forage: NP

Description: It is an erect annual herb. Leaves are opposite below, alternate above, pinnately compound with sessile, lanceolate leaflets. Flowers are arranged in terminal solitary heads. Ray flowers 5–10 or in multiple series; disc flowers are numerous. Fruits are black, cylindrical achenes.

Toona ciliata M. Roem.

Syn: *Cedrela toona* Roxb. ex Rottler & Willd.; *C. australis* Mudie; *C. australis* F. Muell.; *C. hexandra* Wall.

Family: Meliaceae

Common Name: Indian Mahogany, Red Cedar

Distribution: It is native to tropical Asia.

Flowering & Fruiting Season: March–May

Source of Bee Forage: N

Description: It is a large tree. Leaves are paripinnate, alternate, leaflets 12–30, opposite or subopposite. Flowers are bisexual, white in color arranged in drooping panicles. Fruit is an oblong capsule with winged seeds.

Trifolium alexandrinum L.

Syn: *Trifolium albiceps* Ehrenb. ex Sweet; *T. alexandrinum* subsp. *serotinum* (Zohary & Lerner) P. Silva

Family: Fabaceae

Common Name: Berseem, Egyptian clover

Distribution: It is native to Egypt, Iran, Iraq, Pakistan, and Gulf States. It is generally cultivated as fodder crop.

Flowering & Fruiting Season: March–May

Source of Bee Forage: NP

Description: It is an erect annual herb with hollow stem. Leaves are alternate with pubescent leaflets. Flowers are arranged in clustered heads. Fruit is a single-seeded pod.

Vachellia nilotica **(L.) P.J.H. Hurter & Mabb.**

Syn: *Acacia arabica* var. *nilotica* (L.) Benth.; *Acacia nilotica* (L.) Willd. ex Delile; *Mimosa nilotica* L.
Family: Fabaceae
Common Name: Gum Arabic Tree
Distribution: It is generally distributed in tropical and subtropical regions of Asia, Australia, and Africa.
Flowering & Fruiting Season: May–July
Source of Bee Forage: NP
Description: It is an evergreen medium-sized tree. Leaves are bipinnately compound, alternate, stipulate. Flowers many are yellow in color arranged in axillary cymose head. Fruits are long flattened pods.

Vitex negundo **L.**

Syn: *Agnus-castus negundo* (L.) Carriere; *Vitex agnus-castus* var. *negundo* (L.) Kuntze
Family: Lamiaceae
Common Name: Bana
Distribution: It is generally found growing in wastelands or near the agricultural fields.
Flowering & Fruiting Season: May–August
Source of Bee Forage: NP
Description: It is a shrub or small tree having quadrangular stem and branches. Leaves are petiolate, opposite, 3–5 foliolate. Flowers are bisexual, campanulate, generally forming large, terminal panicles. Fruit is a succulent drupe.

Vitis vinifera **L.**

Syn: *Cissus vinifera* (L.) Kuntze; *Palatina oblonga* Bronner; *P. sylvestris* Bronner; *Vitis cylindrica* Raf.; *V. densiflora* A. Sav.
Family: Vitaceae
Common Name: Grape
Distribution: Cultivated. It is native to Europe, Western Asia, and Northern Africa.
Flowering & Fruiting Season: March–April
Source of Bee Forage: NP
Description: It is a deciduous vine. Leaves are alternate, simple, palmately lobed. Bisexual flowers arise in dense thyrses. Fruit is an ovoid berry.

Zea mays **L.**

Syn: *Zea segetalis* Salisb.; *Mays americana* Baumg.; *M. zea* Gaertn.
Family: Poaceae
Common Name: Maize
Distribution: Cultivated. It is native to Mexico.
Flowering & Fruiting Season: July–August
Source of Bee Forage: P

Description: It is an erect annual herb having a main culm with nodes and internodes. Leaves are linear to lanceolate. Male and female flowers are produced on the same plant. Male flower called tassel arise in panicles and female flower called ear arises in modified spikes.
Ziziphus mauritiana **Lam.**
Syn: *Ziziphus mauritiana* var. *fruticosa* (Haines) Sebastine & A.N. Henry; *Z. mauritiana* var. *pubescens* Bhandari & Bhansali
Family: Rhamnaceae
Common Name: Indian Jujube
Distribution: It is distributed in India, Pakistan, China, Ceylon, Afghanistan, Africa, and Australia.
Flowering & Fruiting Season: July–September
Source of Bee Forage: NP
Description: It is a tree or large shrub with spreading branches and long stipular spines. Leaves are ovate, oblong, glabrous above, and tomentose beneath. Flowers are greenish-yellow, arranged in axillary cymes. Fruits are globose to ovoid.

4.5 Honey Flow Sources

Honey flow source is the flora having dense plantation secreting large quantity of nectar. Honey flow indicates the availability of abundant sources of nectar along with favorable weather so that the bees have access to forage for that nectar. The two basic factors for honey flow include access to nectar and favorable weather conditions. It is essential that the colonies should be established entirely to avail honey flow. The beekeepers have to select the regions for their hives which are already having flow or they should shift their hives to the places having honey flow. Common honey flow sources include species of *Acacia* spp. (Acacia), *Brassica campestris* (Mustard), *Citrus* spp. (Citrus), *Dalbergia sissoo* (Shisham), *Eucalyptus* spp. (Eucalyptus), *Fagopyrum esculentum* (Buckwheat), *Gossypium* spp. (Cotton), *Helianthus annuus* (Sunflower), *Litchi chinensis* (Litchi), *Ocimum tenuiflorum* (Holy Basil), *Plectranthus* spp. (Spur flower), *Prunus cerasoides* (Himalayan Cherry), *Syzygium cumini* (Jamun), *Toona ciliata* (Toon), *Trifolium alexandrinum* (Berseem), etc.

Chapter 5
Herbs Honey Infusion Methods

Honey and herbs both have immense roles in the folk medicinal system. Both of these have been used either alone or in combination by various ethnic communities all over the world. These are given in decoction, powder, as well as whole form for curing minor infections, wound healing, and immunity boosting.

Honey and herbs are given to affected/diseased persons by simply mixing them in different ratios. Very few studies have been carried out to prepare herb-infused honey under laboratory conditions.

5.1 Methods of Making Herb-Infused Honey

Some of the infusion methods adopted by previous researchers are as follows:

5.1.1 Infusion Method I (Grabek-Lejko et al. 2022)

Blackberries (*Rubus fruticosus*) and raspberries (*Rubus idaeus*) fruits and leaves from two samples each were used. The first was commercially accessible and purchased in dry form from a nearby health shop, and another one was harvested in 2020 from crops in southeast Poland. At the ideal ripening time, fruits were lyophilized, and fresh leaves were kept in shady conditions till drying. The leaves after complete drying were finally grinded using a laboratory mill to prepare fine powder. During the 2020 beekeeping season, rape honey was purchased from an apiary situated in the Podkarpackie area in Poland. The honey thus obtained was processed, heated, and liquified at 45 °C for 48 h in a laboratory.

5.1.1.1 Preparation of Rubus-Enriched Honey

As described by Tomczyk et al. (2020), rape honey was enhanced with fruits and leaves. Liquefied honey was mixed for 60 s four times a day with crystallized honey (99:1) to begin the crystallization process. Then, 1% and 4% of the fruits were concentrated in honey while 0.5% and 1% (w/w) of blackberry and raspberry leaves (powder) were added. The material was mixed again for about 1 min and then set aside at a cool temperature, i.e., 4 °C for 3 days while being mixed two times daily. There were two technical replications for each variety, yielding a total of 32 test samples. Rape honey was used as a control. Honey mixtures were kept for crystallization for about 1 month at room temperature (20 °C) before being tested.

5.1.1.2 Sample Preparation for Analysis

Aqueous ethanol (50%) was extracted using 2.5 g of powdered leaves and fruits using an ultrasound-assisted procedure. Before analysis, the extracts of herbs were passed through filter paper for proper filtration and kept at a cool temperature using the refrigerator. Extracts were lyophilized using lyophilizer for antibacterial and antiviral activities, and the dry mass thus obtained was put off in 5% dimethyl sulfoxide before being pasteurized.

Honey extracts were also prepared so that the same can be infused well and analyzed for different parameters. For this, 10 mL of distilled water was used to dissolve 0.5 g of honey to create 5% solutions. The solid-phase extraction method was used to prepare honey samples for chromatographic analysis. In a nutshell, acidified water of about pH 2.0 was used to dissolve 20 g of each honey before it was run through a cartridge that had been pretreated with methyl alcohol and acidic water. Acidified water was used to elute the sugars, and methanol (2.5 mL) was used to elute the polyphenolic extract.

The antibacterial property of infused honey was evaluated against S. aureus and E. coli, whereas the antiviral activity was examined against a coronavirus surrogate, i.e., bacteriophage phi 6.

5.1.2 Infusion Method II (Putri et al. 2022)

5.1.2.1 Making Sungkai Leaf Simplicia

In Indonesia, the leaves of Sungkai plant were used to enhance honey. The leaves were harvested from young sections (leaf shoots) of mature plants. It has already been reported that young Sungkai leaf extract contains a number of active compounds, including peronemin, sitosterol, isopropanol, phytol, diterpenoid, and flavonoid. These components might contribute to an increase in leukocytes, which are immune system-related body cells. The first step in making Sungkai leaf simplicial was to process young leaves, wash them under running water, drain them, and then dry them for three to 7 days without exposure to sunlight. After that, the dried leaves were powdered in a blender or herbal powder mill. Drying was primarily done to lower the material's moisture, which can prevent the growth of unwelcome bacteria that could degrade the product's quality. The ideal temperature range for drying simplicia materials is between 30 and 60 °C, with 60 °C being the sweet spot (Yamin et al. 2017).

5.1.2.2 Making Sungkai Leaf Extract

The maceration method was used to prepare Sungkai leaf extract. The complete mixing was ensured by stirring leaf powder and solvent.

The solution was filtered multiple times to obtain clear filtrate. The remaining material was repressed for 24 h using a solvent solution that was three times as heavy as the remaining material. It was then filtered and the gathered filtrate was concentrated in a rotating vacuum evaporator at 40 °C to produce a thick extract, and it was then evaporated in a water bath at 60–80 °C.

5.1.2.3 Making Herbal Honey

The infused honey was prepared using different combinations of honey and Sungkai leaf extract. All the formulations were homogenized manually in stainless steel containers. The homogenized mixture was placed in bottles and kept chilled. It is better to store honey at a low temperature than at room temperature because the humidity at room temperature makes it easier for honey to absorb water, which makes fermentation more likely. After critical analysis and evaluation of infused honey, it was concluded that the composition of the herbal honey greatly influences the flavor and scent that is created.

5.1.3 *Infusion Method III (Ewnetu et al. 2014)*

5.1.3.1 Aqueous Garlic Extract Preparation

In Georgetown (USA), developed fresh garlic bulbs were taken for their infusion in honey. The bulbs were washed and peeled properly to remove impurities to obtain 100% extract. The extract was further diluted, resulting in 75%, 50%, and 25% concentrations. The volume of the extract in each concentration was made to be 10 mL. Each concentration's aqueous extracts were kept in individual glass vials with tight-fitting closures and labels. The Whatman discs were punched, and put into each vial, and sterilization was carried out using an autoclave for about 15 min. After that, the glass vials were kept chilled at 4 °C (Ewnetu et al. 2014).

5.1.3.2 Aqueous Ginger Extract Preparation

Ginger rhizomes were procured from the local market of Georgetown (USA). The rhizomes of ginger were cleaned with fresh water and given an hour for drying naturally. The exterior coating was peeled off, and rinsed with water, and then the following procedure was used to extricate it: Rhizomes of ginger were grated. The sterile cheesecloth was then used to filter the ginger that had been grated. The juice that was produced was regarded as the extract's purest form. To create varying concentrations, the extract was further diluted, resulting in 75%, 50%, and 25% concentrations. The volume of each extract concentration was made to be 10 mL. Each concentration's aqueous extracts were kept in their own glass vials with tight-fitting closures and labels. The Whatman discs were punched with a 5 mm paper puncher and kept in glass vials after sterilization. After that, the extract was kept under cool conditions, i.e., 4 °C in the refrigerator (Ewnetu et al. 2014).

5.1.3.3 Filtering and Honey Diluting

Honey samples were collected from Guyana. To produce 100% pure honey, the honey sample was processed and filtration was carried out using fine mesh so that all impurities may be removed. In order to create different concentrations of honey, including 75%, 50%, and 25%, the honey was further diluted using distilled water. The volume of each extract concentration was made to be 10 mL. The rest of the protocol was the same as that for garlic/ginger extract.

5.1.3.4 Making a Honey-Garlic, Honey-Ginger Mixture

The honey-garlic combo was made using only pure honey and garlic/ginger extracts prepared as per the method mentioned above. The ingredients were mixed in 50:50 ratios. In a test tube, honey and garlic/ginger extract were combined to create a honey-garlic stock. Then, using a stirring stick, the honey and garlic extracts were blended together. Four different honey-garlic/ginger combinations were formulated from the honey-garlic stock: 100%, 75%, 50%, and 25%. Each formulation of the mixture's aqueous extracts was kept separately in glass vials with tight-fitting covers and labels and stored at a cool temperature of 4 °C after proper sterilization using an autoclave. The antibacterial potential of honey-garlic and honey-ginger mixtures against *P. aeruginosa* and *K. pneumoniae*, two clinically significant bacteria commonly linked to acute wound infections, were compared during this study.

5.1.4 Infusion Method IV (Tomczyk et al. 2020)

Herbal honey was prepared using mulberry fruit and leaves (*Morus bombycis* and *Morus alba*) in the Rzeszów region, Poland. The fresh leaves of plants were taken and kept for drying in shady areas where there is no direct sunlight. The fruits were then grinded and lyophilized. Using a typical pressure of 0.5 bars and 30 °C temperature, dehydration was conducted for 48 h.

5.1.4.1 Preparation of Plant Extract for Spectrophotometric Assays

The extract of finely ground dried leaves or freeze-dried fruits was taken and exposed to 212 °C for an hour. The extracts were then filtered and kept in a refrigerator at 21 °C till further use.

5.1.4.2 Mulberry Fruits and Leaves Added to Liquid Honey

Mulberry leaves of two different species after drying and freeze-dried fruits were mixed to rape honey that was liquefied at a concentration of 4% (w/w). A suitable quantity of finely chopped samples was added to glass jars of honey. A kitchen mixer was then used to properly mix the entire combination for 60 s. Such prepared samples were kept at 2 °C for 30 days before analysis, out of the sun. The sample was uniformed right before analysis in the event of delamination.

5.1.4.3 Adding Mulberry Fruits and Leaves to the Creaming of the Honey

To begin the crystallization process, liquefied rape honey was mixed with 99:1 g of crystalized honey and stirred using a mixer grinder for 60 s four times a day. Then, honey was added to powdered mulberry leaves or fruits in proportions of 1%, 2%, and 4% (w/w), and the entire concoction was again combined for 60 s in the kitchen mixer. To ensure an equal distribution of additives, these prepared samples were mixed twice daily and kept under cool conditions at 4 °C for 3 days. The samples were kept at 212 °C without sunshine for 30 days after full crystallization before being analyzed. This experiment was carried out to find out the antioxidant properties, phenolic profile, and diastase enzyme activity.

5.1.5 Infusion Method V (Miłek et al. 2023)

Fresh rapeseed honey was acquired from an apiary situated in the Subcarpathia region, Poland. Freeze-dried black raspberry (*Rubus occidentalis*), blackcurrant (*Ribes nigrum*), wild rose hips (*Rosa canina*), barberry (Berberis vulgaris), as well as dried elderberry flower (Sambucus nigra) were procured from the local market. Dried sea buckthorn leaves (*Hippophae rhamnoides*) and micronized dried apple powder (Malus domestica) were purchased from The BiGrim Company (Wojciechów, Poland).

5.1.5.1 Preparation of Samples of Enriched Honey

Honey was initially decrystallized by heating at about 42 °C for 48 h in the laboratory. The honey was then blended with two pulverized plant additions (2% and 4%, w/w). The samples were mixed twice daily and kept at 4 °C for 3 days. After the crystallization process, the infused honey samples were next exposed to examination. After 90 days of storage, the analysis was carried out again.

The study's objective was to assess the effect of adding particular fruits and herbs from the "superfoods" category on honey and its bioactivities. Infused honey was examined for phenols, flavonoids, antioxidants, and antibacterial activity against four strains of Gram-positive (*Staphylococcus aureus*) and Gram-negative (*Escherichia coli* ATCC, *Pseudomonas aeruginosa*, and *Klebsiella pneumoniae*) bacteria.

5.1.6 Infusion Method VI (Ewnetu et al. 2014)

5.1.6.1 Making Solutions Using Honey and Ginger Extract

The rhizomes of ginger were taken and cleaned with fresh water before being cut into small pieces and kept for drying overnight in a micro-oven at 37 °C. Dried ginger pieces were grinded using a grinder to make a fine powder of it. Ginger extracts were made by combining 20 g of ginger powder with 100 mL of methanol and ethanol. One milliliter of ginger extract was dissolved in one milliliter of water to create a 50% ginger solution (50% v/v). Ginger powder mixed in water was used as the negative control. Honey-ginger infusion was made by combining 1 mL of ginger extract with 1 mL of honey extract. This mixture was then diluted in 2 mL of distilled water to make a 50% v/v honey-ginger extract solution. The antimicrobial properties of blends of ginger-infused honey were assessed against *Klebsiella pneumoniae* (R), Escherichia coli, and *Staphylococcus aureus*.

5.1.7 Infusion Method VII (Jafari et al. 2023)

5.1.7.1 Plant Materials and Processing

In Iran, herbal honey was prepared in a laboratory using Black cardamom and Zataria multiflora. The leaves of both plants were grinded using an electric grinder. The herbal powders were divided into individual Erlenmeyer flasks, and each one was added with 100 mL of 96% ethanol. The mixture was shaken at a temperature of 37–40 °C for about 48 h and 200 rpm to extract the plant's entire active component into the solvent. The content was then filtered using Whatman's filter paper (Whatman, UK). The filtrate was then centrifuged for 10 min at 3000 rpm. Finally, it was stored in the refrigerator till further use.

The antimicrobial activity of herbal honey prepared was analyzed against *S. aureus*, *E. coli*, and *P. aeruginosa*. Additionally, the cytotoxic effects of herbal honey on human erythrocytes and renal epithelial cells were also analyzed.

5.1.8 Infusion Method VIII (Laksemi et al. 2023)

5.1.8.1 Making Extracts

In this study, Citrus aurantifolia acquired through the local market in Bali, Indonesia, was utilized to prepare herbal honey. The fruit was first washed under running water before being chopped, blended, and weighed up to 500 g in the extraction process. The samples were dissolved in a solvent at a 1:2 ratio in order to proceed with the next stage of maceration. After agitating, the mixture was allowed to stabilize for a full day. To extract the extract and discard the dregs, the crude extract was filtered. Evaporation was the following stage. For 1.5 h, the crude extract was rotated at 400 °C and 175 bar of pressure in order to produce a profuse extract. This extract was kept in a sealed, dark container. Until the time came to use the samples for research, storage was done at low temperatures (in a freezer or refrigerator). Citrus aurantifolia 100 mg/kg BW was the dosage utilized in this investigation. The impact of herbal honey prepared was tested on rat model for antimalarial properties.

Chapter 6
Herbal-Infused Honey vis-à-vis Human Health

6.1 Introduction

"The knowledge, skills, and practices based on the ideas, beliefs, and experiences unique to different cultures, employed in the prevention, diagnosis, improvement or treatment of physical and mental disease" refers to traditional medicine (World Health Organization, http://www.who.int/topics/traditional_medicine/en/), and the use of herbs is a core part of all systems of traditional medicine (Rishton 2008; Schmidt et al. 2008).

Around the world, the use of herbal remedies to treat a variety of medical conditions is still growing quickly. Natural remedies are seeing a huge upsurge in popularity and public interest in both developed and developing nations. India possesses an extensive traditional medical system. The majority of medicinal practices, including Ayurveda, Unani, homeopathy, Sidha, etc., rely on plants (Chaughule and Barve 2023). Historically, the earliest medications utilized in the traditional medical systems of several nations, and civilizations were made from herbs, which are broadly defined as any type of plant or plant product as well as plant extracts. Medicinal plants and herbs have long been a frequent source, either as pure active components or as traditional extracts (Fabricant and Farnsworth 2001; Tachjian et al. 2010). Famous medications from herbal and plant origins include aspirin from the *Salix alba* L. tree, digoxin (cardiac glycoside) from *Digitalis purpurea*, ephedrine from *Ephedra sinica*, lovastatin from *Monascus purpureus* L., taxol from *Taxus brevifolia*, reserpine from *Rauvolfia serpentina*, and many more (Frishman et al. 2009; Cragg and Newman 2013). Interestingly, reserpine remains an effective medication for hypertension (Weber et al. 2014). The discovery of antimalarial medications, quinine from the bark of *Cinchona* species and artemisinin from *Artemisia annua* L., is a classic illustration of how ethnomedicine may assist drug development (Cragg and Newman 2013).

© The Author(s), under exclusive license to Springer Nature Singapore Pte Ltd. 2024
R. Kumar et al., *Biomedical Perspectives of Herbal Honey*,
https://doi.org/10.1007/978-981-97-1529-9_6

6.1.1 Herbal Therapy for Disease Treatment

Pathological illnesses include a wide spectrum of ailments that affect numerous organs and systems in the body, and treatment options vary depending on the disease and its underlying causes. Despite the broad range of possible therapies for pathological disorders, there are certain limits to these techniques.

Global demand for herbal medicines is rising despite contemporary medical and technological developments. Natural herbal items are increasingly being used to generate contemporary medications, nutritional supplements, cosmetics, and food and beverage additives. These items include naturally occurring chemicals with antiviral, antibacterial, antiprotozoal, and antioxidant activities. The increase in medication resistance has led to a spike in the use of herbal medicine when conventional therapies fail. They are frequently used in conjunction, combined with honey or alone to treat a variety of diseases. Modern versions of these drugs are available in the form of capsules, pills, powders, and granules (Kumar et al. 2024). Herbal and plant treatments are not only inexpensive, but they also contain thousands of bioactive components with established medicinal uses (Pan et al. 2013). The current rise of popular interest in herbal medicines includes the erroneous idea that herbal products are superior and safer than synthetic products (Ekor 2014). Another reason for the renewed interest in natural products is that their biological activity and structural variety are unrivalled by any existing synthetic drug screening library (Davison and Brimble 2019) (Table 6.1).

6.1.2 Herbal Medicine Use in COVID-19 Management

Herbal medicine and its bioactive components have the potential to be effective in COVID-19 prevention and support interventions. Several beneficial herbal medicines can interfere with COVID-19 pathogenesis by decreasing SARS-CoV-2 reproduction and entrance into host cells. Different plant biochemicals are the most desirable herbal drink or fruit that may be introduced as efficient adjuvant components in COVID-19 care, as well as to lower fever and cough, which are the most prevalent COVID-19 complications, due to their anti-inflammatory effects. Herbal items such as *Gymnanthemum amygdalinum, Azadirachta indica, Nigella sativa,* and *Eurycoma longifolia* can be utilized. Herbal medications including *Glycyrrhiza glabra, Thymus vulgaris, Allium sativum, Althea officinalis,* and *ginseng* may help prevent and treat COVID-19 (Demeke et al. 2021). It is critical to advance the clinical development of herbal remedies in order to determine their therapeutic use in treating a variety of illnesses, including COVID-19, as demand for these medicines rises (Onyeaghala et al. 2023).

6.1 Introduction

Table 6.1 Most popular herbal medicines, including their main active compounds and medicinal uses

S. no.	Herbal medicine	Active compound	Medicinal uses	References
1	Echinacea	Echinacoside, Echinacea	Heal many conditions, including cuts, burns, toothaches, sore throats, and upset stomachs	Hostettmann (2003)
2	Ginkgo biloba leaves extract (EGb)	Ginkgolide, bilobalide, quercetin, kaempferol	It is widely used in various degenerative diseases such as cerebrovascular disease, Alzheimer's disease, macroangiopathy, and more	Zuo et al. (2017), Noor-E-Tabassum et al. (2022)
3	Turmeric (*Curcuma longa*)	Curcumin	Numerous illnesses, such as chronic inflammation, pain, metabolic syndrome, and anxiety, may be treated by it	Hewlings and Kalman (2017)
4	Valerian root	Valerenic acid, sesquiterpenes, iridoid glycosides	Used to treat anxiety and sleeplessness as well as to reduce tremors, headaches, and palpitations in addition to restlessness	Nunes and Sousa (2011)
5	Ginger	6-shogaol and 6-gingerol	Its most well-established contemporary uses include treating colds, nausea, migraines, high blood pressure, and pregnancy-related nausea	Aggarwal et al. (2008), Nicoll and Henein (2009)
6	Chamomile	Apigenin, quercetin, rutin, luteolin	used as a treatment for upper respiratory infections, wounds, nausea, diarrhea, constipation, and stomach discomfort	Singh et al. (2011)
7	*Hypericum perforatum* (St. John's Wort)	Hypericin, hyperforin, and flavonoids (quercetin, rutin, kaempferol)	It was used to treat depression, sleeplessness, wound healing, and a number of renal and lung conditions earlier. These days, most prescriptions for it are for mild to severe depression	Ng et al. (2017), Mullaicharam and Halligudi (2019)
8	Ginseng	Ginsenosides	Build immunity, regulate blood sugar, improve focus, reduce inflammation, and also treat cancer and cardiovascular diseases	Wee and Chung (2012), Shaito et al. (2020)
9	Elderberry (*Sambucus nigra*)	Polyphenols and anthocyanins	It has been used for a very long time to treat constipation, colds, viral infections, headaches, and nerve discomfort	Fossum (2014)

6.2 Herbal Infusions: Harmony in Therapeutic Potential

In light of increasing antibiotic resistance, the quest for alternative medicines, including herbal antibiotics, has become critical, particularly for bacterial infections. The bioactive ingredients work together to maximize therapeutic efficacy while reducing side effects. This integrated method might be used as an alternative to traditional medicinal therapies. Herbal infusions are an easily consumed form of plants. Herbal infusions have long been used in traditional medicine and are a popular global beverage option (Poswal et al. 2019). Evidence suggests that bio-actives found in herbal infusions may have a wide range of biological effects, including potential antibacterial, antioxidant, anti-inflammatory, antiallergic, antithrombotic, and vasodilatory actions, as well as antimutagenic, anticarcinogenic, and antiaging effects (Chandrasekara and Shahidi 2018).

Several ethnobotanical studies have indicated that a variety of herbal species are frequently mixed in order to increase their pharmacological effects. Often, a mixture of phytochemicals is responsible for the antioxidant activity, which is due to either additive and/or synergistic effects (Guimarães et al. 2011).

6.2.1 Herbs Infused with Honey

There is an increasing interest in whether herbal infusions might help healthy living and preventative health. Honey and herbs have long been renowned for their health advantages. Honey's particular chemical makeup, which includes vitamins, minerals, and enzymes, makes it a great substrate for herbal infusion. The presence of physiologically active chemicals, such as plant secondary metabolites like phenolic acids, flavonoids, coumarins, tannins, lignans, and terpenoids contributes to their pro-health qualities (Wink 2015). The investigation of honey infused with herbs might provide a viable therapeutic intervention against the numerous common terrible illnesses. So much work goes into combining the medicinal advantages of herbs and honey to enhance their activity.

Herb-infused honey is a term used to describe a variety of new products that have been produced by infusing honey with the medicinal elements of herbs (Dżugan et al. 2017). Another well-liked product which is made by providing bees with a sugar medium enhanced with fruit juices or herbal extract is herb honey. Compared to nectar honey, herbal honey has more health-promoting qualities, including antibacterial and antioxidant activity (Isidorov et al. 2015; Lukasiewicz et al. 2015). Such products can contain chemicals that are not provided in flower nectar derived from plant roots or leaves. *Melilot* honey is one such example which is produced by bees from the nectar of two different species of sweet clover: yellow sweet clover (*Melilotus officinalis*) and white sweet clover (*Melilotus albus*). Melilotus honey is thought to include coumarin chemicals, which are known to have antioxidant and antibacterial properties (Sowa et al. 2017). Honey itself has a lot of health benefits and serves as the best choice for herbal infusions.

6.2.2 Health Benefits of Honey

Honey has been utilized for medicinal and nutritive purposes since antiquity. The antioxidant activity of honey is important as a hepatoprotective and cardioprotective agent (Biluca et al. 2020). Additionally, honey protects against gastrointestinal illnesses (Rao et al. 2016). Honey possesses anti-inflammatory (Biluca et al. 2020; El-Seedi et al. 2022) and anticancer properties against breast, cervical (Fauzi et al. 2011), and prostate cancers (Samarghandian et al. 2011) as well as osteosarcoma (Ghashm et al. 2010). Furthermore, honey is traditionally utilized as an antidiabetic, antioxidant (Ahmed et al. 2018) hypolipidemic medication (Adnan et al. 2011) and helps in thyroid disorders. A number of gram-negative bacteria, such as *Escherichia coli, Salmonella sp., Shigella sp., Helicobacter pylori,* and others are significantly or extremely sensitive to the biologically active compounds in honey. All these health benefits are because of the active components present in honey.

6.2.2.1 Biochemical Composition of Honey

According to its chemical makeup, honey includes roughly 200 bioactive chemicals (carbohydrates, amino acids, proteins, enzymes, carotenoids, minerals, vitamins, polyphenols, *etc.*) as mentioned in Fig. 6.1, which are responsible for its various medicinal properties described above (Chew et al. 2018; Mračević et al. 2020).

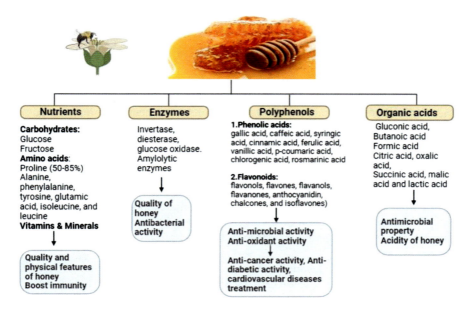

Fig. 6.1 Biochemical composition of honey and their respective biological activities

Numerous variables, such as the honey's geographical origin, floral source, weather conditions, any treatments used and seasonality, affect how much of these components it contains (Clearwater et al. 2018).

Honey is a good source of both macro and micronutrients, amino acids, and vitamins. Higher amounts of glucose and fructose are present and are primarily responsible for the energy value and physical features of honey, including hygroscopicity, granulation, and viscosity. Carbohydrates, including mono- and disaccharides, make up 95% of its dry weight. Amino acid supplementation, particularly of cysteine, glutamine, and arginine, is proven to boost immune system function. Additionally, honey contains trace levels of minerals and vitamins (ascorbic acid, riboflavin, folic acid, niacin, pantothenic acid, and vitamin B6) (calcium, potassium, iron, zinc, phosphorus, magnesium, manganese, selenium, and chromium) (Gośliński et al. 2021). Organic acids (gluconic, succinic, acetic, butyric, citric, lactic, and malic acid) give honey its taste, acidity as well as antibacterial properties. Organic acids like lactic, formic, and oxalic have been shown to be useful in treating ectoparasitic mites.

A percentage of the proteins found in honey are enzymes derived from pollen, nectar, and bees. Invertase and glucose oxidase are the main enzymes. Amylolytic enzymes, such as α-amylases and β-amylases, hydrolyze starch chains to create maltose and dextrin, respectively. The activity of these enzymes is one of the key elements in determining the quality of honey. The primary acid in honey, gluconic acid, is produced when the enzyme glucose oxidase converts glucose to δ-gluconolactone. Hydrolysis of this molecule yields hydrogen peroxide (H_2O_2) which provides honey its antibacterial qualities.

Phenolic compounds: The chemical class known as polyphenols is diverse and may be further classified into two groups: non-flavonoids and flavonoids (flavanols, flavones, flavanols, flavanones, anthocyanidin, chalcones, and isoflavones) (phenolic acids). These compounds have been identified as the primary cause of honey's antioxidant activity, which is primarily linked to its capacity to scavenge free radicals by forming less harmful and more stable molecules. When free radicals release hydrogen from one of their hydroxyl groups, phenolic compounds neutralize them; the quantity of hydroxyl groups in phenolic compounds determines how active they are (Cianciosi et al. 2018). Honey is assumed to have much of its biological action coming from its phenolic and flavonoid compounds. It has been shown that a variety of Gram-positive and Gram-negative bacteria are inhibited by polyphenols present in floral nectar. Cinnamic acid, gallic acid, caffeic acid, ferulic acid, vanillic acid, *p*-coumaric acid, chlorogenic acid, syringic acid, rosmarinic acid, and their derivatives are among the many phenolic acids found in honey. Among the substances found in honey with antibacterial properties are benzoic, syringic, cinnamic, ferulic, and caffeic acids (Feknous and Boumendjel 2022).

Four categories of flavonoids may be found in honey: flavonols (quercetin, kaempferol, galangin, myricetin, and fisetin), flavanones (naringin, hesperidin, naringenin, pinobanksin, and pinocembrin), flavones (apigenin, luteolin, tricetin, genkwanin, wogonin, and acacetin), and tannins (ellagic acid). Propolis is often the source of flavonoids found in honey (Šedík et al. 2019). Kaempferol, pinobanksin, chrysin, quercetin, galangin, and pinocembrin are the antimicrobial compounds (Couquet et al. 2013).

6.2.3 Honey in Herbal Medicine Formulations

Besides having numerous biological activities like antioxidant and antimicrobial, *etc.* which in synergism with herbs enhance the efficiency of herbal treatments, honey also provides other benefits like helping in the storage of herbal medicines, and can be used as a carrier for medicinal herbs. Honey draws other therapeutic chemicals deeply into the tissues by acting as a catalytic carrier, or anupana. Honey facilitates heart-healthy herbs' quick absorption into the bloodstream (Johari 1994). Traditional formula may be a safe and effective supplemental medicine for the treatment. Compound honey syrup (CHS) is made of a combination of honey and the extracts of five herbal medicines: galangal, cardamom, saffron, ginger, and cinnamon. Each of these medicines has anti-inflammatory, antibacterial, antitussive, bronchodilator, and anticholinergic properties and is used to treat lung diseases (Poursaleh et al. 2022). A new herbal paste formulation comprising turmeric, tulsi (holy basil), and honey has been compared in several trials for its efficacy in treating Oral Sub-Mucous Fibrosis (OSMF) (Mobeen et al. 2023).

6.3 Study of Synergism of Honey with Various Herbs

Synergism, as used in medicine, is the outcome of combining two or more ingredients to create a superior product with the greatest qualities. Synergism is one of the main factors observed to make herbs more efficient after infusion with honey (Spoială et al. 2022). Infusing honey with herbs, spices, or fruits may increase its health advantages. For example, adding turmeric to honey may improve its anti-inflammatory characteristics, while adding lemon to honey may increase its immune-boosting benefits. Seven plant species have been found to work well with honey. These species include *Phyllanthus emblica L., Euphorbia hirta L., Nigella sativa L., Curcuma xanthorrhiza Roxb, Capparis spinosa L., and Allium sativum L.* Each species has been shown to have special medicinal properties when combined with honey to treat a range of illnesses.

6.3.1 Allium sativum L

Garlic, also known as *Allium sativum L.*, is a member of the Alliaceae family and is widely used as a spice and a well-liked remedy for a variety of diseases and physiological problems. (Takhtadzhian 1997).

6.3.1.1 Biological Activity

The bulbs of *Allium sativum* contain compounds containing sulfur, viz. thiosulfates (allicin), vinyldithiins (2- vinyl-(4H) 1, 3-dithiin, 3-vinyl-(4H)-1, 2-dithiin), ajoenes (E-ajoene, Z-ajoene), and sulfides (diallyl disulfide, diallyl trisulfide, and others). Allicin (allyl thiosulfinate) is a sulfenic acid thioester which is having strong antioxidant activity (Diretto et al. 2017).

The method of preparation determines the biological activity of a garlic extract. The major ingredient in the room-temperature extract is allicin, which has a potent antibacterial effect. Small quantities of several additional thiosulfinates and complex sulfinyl components, such as the antithrombotic ajoenes, are also found in addition to allicin (Yoo et al. 2014a, b). The primary ingredient in the aqueous extract of heat-treated garlic is alliin, which has no smell. Garlic, when distilled in steam, produces an oily substance that contains diallyl, methyl allyl, and dimethyl, which are all derived from the thiosulfinates (El-Saber Batiha et al. 2020). Studies have demonstrated the biological benefits of garlic oil, including its anticancer and antioxidant capabilities. Organosulfur compounds, including s-allyl cysteine, s-allyl mercaptocysteine, and many sulfur-containing amino acids, are formed when garlic is exposed to a cold aging process (Yoo et al. 2014a, b). It has been demonstrated that s-allyl cysteine and s-allyl mercaptocysteine are anticarcinogenic and protect against liver damage (Yi et al. 2019).

6.3.1.2 *Allium sativum L* Infused with Honey

Medical professionals as well as traditional practitioners have been using herbal plant medicines for burns and soft tissue wounds for a long time in underdeveloped nations. According to research done on mice, applying garlic aqueous extract topically to wounds considerably speed up the healing process as compared to applying honey alone. Honey is said to provide wound healing qualities that include encouraging the formation of new tissue, moist wound healing, managing fluids, and encouraging epithelialization. Honey inhibits the development of bacteria primarily by acting as a hyperosmolar medium. Honey is hyperosmolar due to its high sugar content. Because of its high viscosity, honey creates physical barriers that keep germs from colonizing wounds and keeps them wet, which seems to be beneficial and speeds up the healing process (Sidik et al. 2006). Honey's antibacterial properties also helped to deodorize the wound and its inflammatory properties lessened the amount of discomfort. Honey's antimicrobial qualities have been linked to hydrogen peroxide, which is created by phenolic chemicals and glucose oxidase, two substances that are found naturally. Honey may have a role as an antioxidant in heat injury because of the enzyme catalase, which is found in honey. Levulose, one of the nutrients found in honey, enhances the availability of local substrates and could aid in the process of epithelialization. It appears that honey's antibacterial qualities speed up the process of epithelialization (Lusby et al. 2002).

For millennia, people have utilized garlic for medical purposes. Allicin is the component of garlic that is most active. Alliinase enzymes are triggered by crushing garlic cloves, and they help transform alliin into allicin (Borlinghaus et al. 2014). Fibroblasts are essential for the healing of wounds. It was proposed that allicin activates fibroblasts, which promotes more ordered and efficient wound healing. Allicin is the primary ingredient in garlic, and it possesses antimicrobial qualities. Garlic's antibacterial properties have been linked to the sulfurides found in garlic oil (Gruhlke et al. 2011). Higher diallyl sulfide concentrations in garlic oil were associated with stronger antibacterial activity. The synergistic effect of honey infused with *allium sativum L.* improves wound healing in animals (Ness et al. 1999).

6.3.2 Alpinia officinarum

Alpinia officinarum, often known as galanga, is a plant that belongs to the Kingdom Plantae and Division Magnoliophyta. Its specific species name is galanga, and it is a member of the Genus Alpinia, Family Zingiberaceae, and Order Zingiberales (Ahmad et al. 2023).

6.3.2.1 Biological Activity

1,8-cineol, α-pinene, β-bisabolene, β-farnesene, α-bergamotene, α-fenchyl acetate, Galago flavonoids, 1′S-1′-acetoxychavicol acetate (ACE), phenylpropanoids, and 1′-acetoxychavicol acetate are the active compounds present in this plant (Verma and Sharma 2022). The three major classes of chemical compounds found in galangal rhizomes are diarylheptanoids, glycosides, and flavonoids. Galangal is said to possess biological actions, such as anticancer, antiulcer, antibacterial, and antifungal qualities (Weerakkody et al. 2011; Malek et al. 2011).

6.3.2.2 *Alpinia officinarum* Infused with Honey

Alpinia officinarum and honey both contain large concentrations of galangin, a flavonoid belonging to the flavonol class (Amal et al. 2018). Honey, propolis, bee glue, and bee wax include distinctive flavonoids called galangin, chrysin, tectochrysin, pinobanksin, and pinocembrin (Tomás-Barberán et al. 1993). It is a significant component in plants and fruits. There is little information available on galangin's antibacterial properties (Ahmad and Beg 2001). The hyphal development of *Gigaspora margarita* revealed the antifungal effects of galangin.

Honey syrups, prepared by combining honey with extracts of *Alpinia galanga*, cardamom, saffron, ginger, and cinnamon were administered to asthmatics and the experimental group. The entire ACQ (Asthma Control Questionnaire) score was assessed, and the outcomes showed that compound honey syrup is a safe and efficient treatment for pediatric asthma and its related symptoms (Choopani et al. 2017).

6.3.3 Capparis spinosa *L*

A member of the Kingdom Plantae's Division Magnoliophyta, *Capparis spinosa L.* is also known as the caper bush or Himsra. Its specific species name is spinosa, and it is a member of the Genus Capparis, Family Capparaceae, and Order Capparales (Shahrajabian et al. 2021).

6.3.3.1 Biological Activity

Capparis spinosa contains a variety of chemical components, including flavonoids, glucosinolates, phenolic acids, terpenoids, and alkaloids. Cappariloside A, stachydrine, hypoxanthine, and uracil in aerial parts are the principal constituents of the aerial parts. Aqueous extracts of *Capparis spinosa* show strong antihyperglycemic and antiobesity properties. Ethanolic extracts of *Capparis spinosa* leaves and roots show suppression of pancreatic α-amylase activity, which may be related to blood sugar regulation. *Capparis spinosa's* ethanolic root bark extract significantly protects against carbon tetrachloride and its associated liver damage in a dose-dependent manner. The aqueous extract of *Capparis spinosa* aerial portion contains p-methyl benzoic acid and has antihepatotoxic properties (Nabavi et al. 2016).

6.3.3.2 *Capparis spinosa L* Infused with Honey

A synergistic interaction between combinations of *Capparis spinosa* leaves and honey in Swiss albino mice exposed to trichloroacetic acid (TCA) shows a mitigating effect on the hepatotoxicity of the substance (Alzergy et al. 2015).

6.3.4 Curcuma xanthorrhiza *Roxb*

Curcuma xanthorrhiza Roxb., or Javanese turmeric, is a member of the Division Magnoliophyta of the Kingdom Plantae. The exact species name of this plant is xanthorrhiza, and it belongs to the Genus Curcuma, Family Zingiberaceae, Order Zingiberales (De Padua et al. 1999).

6.3.4.1 Biological Activity

Curcumin, bisacurone, curlone, α-curcumene, bisacumol, bisacurol, ar-turmerone, xanthorrhizol, β sesquiphellandrene, curzerenone, germacrone, β-curcumene, α-turmerone, β-turmerone, and (π)-curcuhydroquinone are the active ingredients in *Curcuma Xanthorrhiza*. Xanthorrhiza curcuma Roxb, often referred to as temulawak, contains curcuminoids (1-2%) and essential oils, such as xanthorrhizol components (31.9%), ß-curcumene (17.1%), ar-curcumene (13.2%), camphor (5.4%), γ-curcumene (2.6%), (Z)-γ-bisabolene (2.6%), and (E)-ß-farnesene (1.2%) (Zhang et al. 2014). The empirical use of temulawak includes in treatment of stomach complaints, liver disorders, fever and constipation, bloody diarrhea, dysentery, rectal inflammation, hemorrhoids, gastric disorders brought on by cold, infections, skin eruptions, acne vulgaris, eczema, smallpox, and anorexia, as well as to lessen uterine inflammation following childbirth (Parastaka and Nihayati 2019).

6.3.4.2 *Curcuma Xanthorrhiza* Infused with Honey

In both Sprague Dawley (SD) rats and human mammary cancer cells, the combination of *Curcuma Xanthorrhiza* and black cumin with honey shows anticancer activity against dimethylbenzene(a) anthracene-induced cancer. This herb-infused honey formulation not only increases glutathione-S transferase activity and reduces tumor nodule development but also elevates CD4, CD8, CD4, and CD25 cell counts. These findings highlight the immunomodulatory and anticancer properties of the herbal mixture (Hidayati et al. 2023).

6.3.5 Euphorbia hirta L

Euphorbia hirta L., sometimes referred to as Dhudika or the Asthma plant, is a member of the Division Tracheophyta in the Kingdom Plantae. The particular species name for it is hirta, and it belongs to the Order Malpighiales, Family Euphorbiaceae, and Genus Euphorbia (Nadkarni 2007).

6.3.5.1 Biological Activity

The active compounds are euphorbin-A, euphorbin-B, euphorbin-C, euphorbin-D, 2,4,6-tri-O-galloyl-β-D-glucose, 1,3,4,6-tetra-O-galloyl-β-D-glucose, rutin, quercitrin, myricitrin, and gallic acid (Sood et al. 2005). Research has demonstrated that *Euphorbia hirta* possesses antimicrobial, antihelminthic, antiretroviral, antiplasmodial, anti-amoebic, antioxidant, sedative, antispasmodic, antifungal, and antimalaria properties (Kumar et al. 2010a, b). It has been stated that the plant contains flavonoids, polyphenols, phytosterols, alkaloids, and triterpenoids (Basma et al. 2011).

6.3.5.2 *Euphorbia hirta* Infused with Honey

An endemic gastrointestinal condition that is a serious global public health concern is stomach ulcers. An imbalance between the stomach's defensive (bicarbonate and mucus) and aggressive (acid and pepsin) components can lead to stomach ulcers. In one research, the antiulcer and gastroprotective qualities of a crude extract of *Euphorbia hirta* combined with honey were tested on rats. According to this research, the administration of 200 mg/kg of *Euphorbia hirta* crude extract alone resulted in 54% inhibition of ulceration; when combined with honey, this inhibition rose to 94%; however, the inhibition of ulceration caused by honey alone was 89.47%. This suggested that the combination of *Euphorbia hirta* and honey had a synergistic impact that improved the prevention of ulceration, as seen by the stomach mucosa's protection (Onyeka et al. 2020).

6.3.6 Nigella sativa *L.*

Nigella sativa L., sometimes referred to as Kalonji or Black cumin is a member of the Division Magnoliophyta in the Kingdom Plantae. It is categorized as belonging to the Genus *Nigella*, the Family Ranunculaceae, the Order Ranunculales, and the particular species name *sativa* (Pruthi 1976).

6.3.6.1 Biological Activity

Certain phytochemicals such as carvone, nigellone, thymoquinone, thymol, nigellicine, nigellicimine, nigellicimine N-oxide, cholesterol, campesterol, gramisterol, lophenol, sitosterol, etc. are present in the seeds and seed oil of *Nigella sativa*. The herbaceous plant known as black cumin (*Nigella sativa L.*) is widely used in the culinary industry, particularly in baking, and is also used in traditional medicine to cure or prevent a number of illnesses, including dyslipidemia, diarrhea, and asthma. Black cumin seeds include protein, polyphenols, alkaloids, saponins, and essential oils (Mamun and Absar 2018). Extensive research using the seed combined with honey or other natural ingredients has shown strong antiviral, antibacterial, and anti-inflammatory properties (Hassan et al. 2012). The plant is exclusively grown for its seeds. Few researchers have looked at black cumin honey, despite the fact that many have used black cumin seed combined with honey. The dark-colored honey is eaten as a preventative measure for health (Rayhan et al. 2019).

6.3.6.2 *Nigella sativa L* Infused with Honey

A study was carried out by researchers to assess the effectiveness of topical honey treatments based on the anti-inflammatory and antibacterial properties of *Nigella sativa* and honey, as well as their documented therapeutic benefits for promoting wound healing. The use of a topical mixture made with *Nigella sativa* seed oil and honey enhanced and expedited the healing of wounds (Javadi et al. 2018).

6.3.7 Phyllanthus emblica L

Phyllanthus emblica L., sometimes referred to as Indian gooseberry or amla, is a member of the Division Magnoliophyta of the Kingdom Plantae. It belongs to the Genus Phyllanthus, the Family Phyllanthaceae, the Order Malpighiales, and the particular species name is emblica. Indian gooseberry, or amla, is a well-known plant that is valued for its many uses in traditional medicine and cooking (Walia et al. 2015).

6.3.7.1 Biological Activity

Phyllanthus emblica fruits include ellagic acid, fibers, nicotinic acid, gallic acid, carbohydrates, proteins, lipids, minerals, phyllembelin, and phyllembelic acid. The following acids are found in seeds: myristic acid, palmitic acid, stearic acid, oleic acid, and linolenic acid. The alkaloids phyllantidine and phyllantine, as well as chebulic, gallic, ellagic, chebulagic, chebulinic, and amlic acid are present in the leaves. The plant's roots include lupeol and ellagic acid. There is leucodelphinidin in the bark (Mazumder et al. 2023). Astringent, carminative, digestive, stomachic, laxative, alterant, cooling, anodyne, aphrodisiac, rejuvenative, and diuretic are some of the qualities of fruits. Tridosha, diabetes, bronchitis, cough, asthma, ophthalmopathy, cephalalgia, ophthalmopathy, dyspepsia, colic, flatulence, hyperacidity, cardiac problems, gray hair, and other vitiated disorders can all be effectively treated with them (Variya et al. 2016)

6.3.7.2 *Phyllanthus emblica L* Infused with Honey

Gastroesophageal reflux disease (GERD) is a disorder that causes erosion and ulcers in the stomach. An in vitro study with the herbal formulation that included amla and honey along with Pantoprazole and Rebamipide for the treatment of GERD was compared with monotherapy (an herbal preparation containing amla and honey) to determine the herbal formulation's therapeutic efficacy. Research was done in vitro using isolated rat ileum to compare the antispasmodic effects of honey and amla to acetylcholine. Compared to monotherapy, combination treatment had the most therapeutic effectiveness when used with the natural combination of amla and honey. Amla and honey formulation's in vitro antispasmodic efficacy shows a concentration-dependent reduction in the contractility pattern against acetylcholine. Honey and amla can be consumed to ease stomach digestion and aid with GERD (Mazumder et al. 2023).

6.4 Conclusion

Herbal medicines have been used for centuries to treat and prevent a wide range of health issues, including eczema, wounds, skin infections, swelling, aging, mental illness, cancer, asthma, diabetes, jaundice, scabies, venereal diseases, snakebite, gastric ulcers, and many more. Almost every known human civilization has used herbal medicines for these purposes. This is mostly due to the widespread perception that herbal medications are inexpensive, readily available locally, and free of negative effects. The efficiency of herbal medicines in terms of their activity can be enhanced by increasing various active compounds and the integration of biochemically active compounds with herbs might be used as an alternative to traditional medicinal therapies.

Honey has been heavily utilized in these types of remedies as it is the most significant natural substance that is rich in vitamins, minerals, and proteins. The beneficial effects of antioxidant substances found in honey, including proteins, enzymes, amino and organic acids, polyphenols, and carotenoids, but especially flavonoids and phenolic acids, have made honey a popular choice in this respect. Honey perfectly complements the aromatic qualities of plants, and herbal medication may be made more easily and calmingly by preserving medicinal plants in locally produced honey. This chapter highlights the medicinal importance of both herbs and honey, the most important active compounds involved in their medicinal properties. Special attention was paid to the synergistic effects of honey with other natural products containing bioactive compounds that have therapeutic characteristics and can ensure a synergistic result. Moreover, various herbal plants (*Allium sativum, Alpinia officinarum, Capparis spinosa, Curcuma xanthorrhiza, Euphorbia hirta, Nigella sativa, and Phyllanthus emblica*) with antibacterial, antioxidant, and antimicrobial properties can work synergistically with honey for the treatment of numerous diseases were also mentioned in this chapter.

Chapter 7
Biomedical Perspectives of Herbal Honey

Honey has been defined as the sweet, viscous liquid produced by honeybees from the nectar of blossoms or from the secretion of living plants which the bees collect, transform into honey, and store in honeycombs. Honey is a supersaturated solution of sugars. Honey has been used by man in various ways since time immemorial. It was the first and most reliable sweet substance used by human beings as a taste enhancer and source of energy. In India, honey is given to neonatal even before the mother's milk. It is considered "***Amrut***" in Vedas and has been given special status among consumers due to its natural image and its purported health benefits. Unveiling honey's role in metabolic health, it becomes a steadfast ally in combating various issues. Honey is a potential prebiotic due to its remarkable versatility, influencing the gut microbiota. Honey is a proven wound healer, anti-decongestant, immunity booster, and rejuvenator.

Herbs are also used by humans since time immemorial. The therapeutic properties of hundreds of herbs have already been explored, a few of which have been described in Chap. 3. In many parts of the world, herbs are still used by people to manage or cure various ailments, a description has already been given in Chap. 2.

The powerful fusion of honey and herbs takes a central stage in combating infectious diseases. Its joint antidiabetic and hypoglycemic properties may reverse diabetes-induced complications, providing a promising natural alternative. Moreover, herbal honey may prove highly significant in managing lifestyle diseases, presenting a natural and effective approach to improving health outcomes. The physiological and curative role of herbal honey in reducing the pathogenesis of reproductive disorders may open new avenues for holistic well-being.

In this chapter, the role of herbal honey has been discussed in the control/cure of both communicable and non-communicable diseases with the help of possible mechanisms (Fig. 7.1).

List of prominent diseases to be discussed for their possible cure using herbal honey:

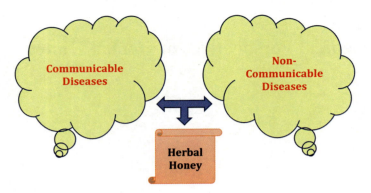

Fig. 7.1 Impact of honey on human diseases

Communicable Diseases:
- COVID-19
- Dermatitis
- Eye diseases
- Hypersensitivity
- Influenza
- Periodontal disease
- Herpes disease
- Viral hepatitis
- Gingivostomatitis
- Anal disorder

Non-Communicable Diseases:
- Diabetes
- Cancer
- Wound management
- Alzheimer's disease
- Gastrointestinal disorders
- Cardiovascular disorders
- Reproductive disorders

7.1 Herbal Honey in Communicable Diseases

7.1.1 Honey and COVID

Coronaviruses (CoVs) are RNA viruses capable of causing infection in animals including human beings, primarily targeting the upper respiratory tract, lungs, and nervous systems. Major causative organisms are SARS-CoV (severe acute respiratory syndrome coronaviruses) and MERS-CoV (Middle East respiratory syndrome coronaviruses), which have been reported to cause infection during the last few decades. Earlier there were no reports of their pathogenicity among human beings; however, huge infections and mortality were caused by SARS-CoV all over the world. It was reported from Wuhan city of China and then reached each and every part of the world leading to 6,955,497 deaths (Source: WHO). The person suffering from SARS-CoV2 reports high fever, cough, headache, fatigue, vomiting, etc.; however, symptoms depend upon age and immunity status of the human body.

Sometimes, there is no specific symptom in the body, still tests positive for COVID. Such a condition is called asymptomatic and usually heals away with the passage of time. In the patient having some or other kind of chronic disease, SARS-CoV2 infection leads to metabolic disruption inside the body such as acute respiratory failure further leading to lung collapse, cardiac arrest, and acute hepatonephric issues. Visceral examination of such patients during postmortem shows complete alveoli damage, edema, and the presence of a hyaline membrane as evidence of severe infection. In very few cases, pneumonia is also reported in patients suffering from SARS-COV2 infection. Both SARS and MERS (members of the beta-coronavirus family) pathogenesis have similar kinds of symptoms (Fig. 7.2). Infected patients exhibit a spike in inflammation as well as weakened immune response, which may trigger a "cytokine storm," "cell apoptosis," "leakage from blood vessels," and "undesired B/T cell responses." All these factors may induce ALI/ARD and in some patients cause death as well. SARS-CoV2 that resulted in

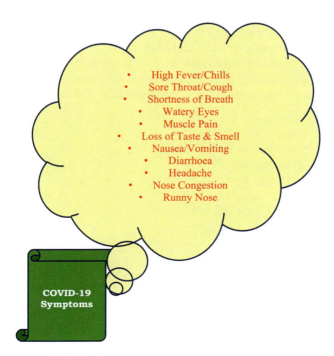

Fig. 7.2 Symptoms of COVID-19

the COVID-19 pandemic reported 79.0%, and 51.8% similarity with SARS-CoV and MERS-CoV on the genetic basis.

In COVID-19-affected patients, excess production of cytokines may lead to severe lung infection. Extensive increase in interleukins IL-1β, IL-4, IL-10, interferon γ, IP-10, and MCP-1 can be seen in serum analysis of SARS-CoV2 infected people in comparison to SARS. Also, low immunity and high inflammation are reported in symptomatic as well as asymptomatic COVID patients due to which cytokines and chemokines are greatly expressed. This leads to further progress in the disease. Interleukin 10 released by T helper cells controls the virus inside the body in general; however, in SARS-CoV2-infected people, extensive increase in IL-10 has been reported which is otherwise supposed to decrease during infection. All these factors lead to programmed cell death.

Herbal honey has many therapeutic properties that prove it is an effective remedy against various pathogens like bacteria and viruses. COVID-19 is a viral disease and highly communicable among the human race. The antiviral, anti-inflammatory, and antioxidative properties of herbal honey prove it best against viral diseases such as COVID-19. Decreased oxidative stress, reducing inflammation, and altering or disrupting the life cycle of SARS-CoV2 may help in combating infection.

Herbs and honey both have been used in traditional medicinal systems to combat respiratory diseases. Tulsi, mint, ginger, and black pepper are generally used along with honey to cure cough and cold.

Oxidative stress during SARS-CoV2 infection leads to several metabolic disorders like inflammation, cardiac arrest, or insulin imbalance. The antioxidant potential of honey inhibits the action of virus by disrupting their replication potential. The presence of polyphenols, carotenes, and other organic acids in herbal honey protects the human body from oxidative DNA damage. It also aids in inhibiting the production of mutagens, gastrointestinal disorders, and inflammation (Fig. 7.3).

Herbal honey helps in the mitigation of infectious diseases like SARS-CoV2 as follows:

(i) Affect viral cell entry
(ii) Alter the life cycle of SARS-CoV2
(iii) Inhibit inflammation
(iv) Increase in oxidants
(v) Decrease oxidative stress
(vi) Inhibit protein binding to ACE-II
(vii) Boosting immunity
(viii) Increase in host defence
(ix) Less susceptible to disease
(x) Inhibit viral proteins

Herbal honey is used to stimulate T cells, B cells, and other defence cells that produce various cytokines like IL-1, IL-6, IFN-γ, and TNF-α. IFN-γ has the potential to attract viral proteins of SARS-CoV2 and finally combat the respective infection. Due to the proliferation of B & T lymphocytes, the adaptive immune response can be induced against SARS-CoV2. Herbal honey also contains a special sugar

7.1 Herbal Honey in Communicable Diseases

Fig. 7.3 Mitigating mechanism of SARS-CoV2 by herbal honey

derivative "nigerose" that stimulates immune response and maintains the formation of immune/defence cells in virus-infected patients. Herbal honey has a profound effect on IgE, hepatic enzymes, and ferritin that has a direct connection with SARS-CoV2 infection (Fig. 7.4).

All these findings conclude that herbal honey has strong potential to fight COVID-19 infections, however, in-depth and more scientific validation is required to explore the significance of herbal honey in the treatment and management of serious pathogens such as SARS-CoV2.

7.1.2 Herbal Honey and Dermatitis

The term "dermatitis," also known as eczema, is a non-infectious and inflammatory skin disorder. Dermatitis comes in various forms, including seborrheic dermatitis (found in both infants and adults), primary irritant dermatitis, allergic contact

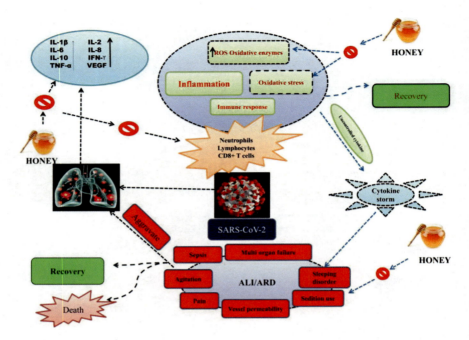

Fig. 7.4 Physiological impacts of herbal honey in SARS-CoV2 infected patients

dermatitis, dermatitis linked to venous hypertension, dyshidrotic dermatitis (also called idiopathic vesicular dermatitis of the hands and feet), photoallergic dermatitis, HIV-associated dermatitis, nummular dermatitis, and atopic dermatitis (AD). Atopic dermatitis is a prevalent condition and stands out as the only type of dermatitis that can affect a person throughout their entire life span. It can be caused by various factors, including allergens, irritants, and genetic predisposition. While numerous treatments are available, herbal honey has gained attention for its potential therapeutic benefits in dermatitis management in recent years, drawing from its historical use in wound healing and its rich chemical composition. Honey's use in wound healing and skin conditions dates back to ancient civilizations, including the Egyptians and Greeks. Ancient civilizations recognized honey's ability to promote tissue repair and reduce inflammation. In traditional medicine, honey was applied to wounds, burns, and skin conditions, showcasing its long-standing cultural significance as a healing agent. With the advent of modern medicine, honey's traditional uses fell into relative obscurity, but recent scientific research has rekindled interest in its potential dermatological benefits. They recognized its ability to promote tissue repair and reduce inflammation. Honey is a complex mixture primarily composed of sugars (glucose and fructose), water, vitamins, minerals, and bioactive compounds. These bioactive compounds include flavonoids and phenolic acids, which contribute to honey's antimicrobial, antioxidant, and anti-inflammatory properties. Such a diverse chemical composition underscores honey's versatility in potential dermatological applications.

7.1.2.1 Causes of Dermatitis

Dermatitis, including various forms like atopic dermatitis, seborrhoeic dermatitis, contact dermatitis, and others, shares some common causes and molecular mechanisms.

7.1.2.2 Allergens and Irritants

Contact dermatitis is frequently triggered by exposure to allergens or irritants. Allergens can include substances like nickel, latex, certain plants (e.g., poison ivy), and cosmetics. Irritants may include harsh soaps, detergents, and chemicals. These compounds can cause skin irritation when they encounter skin, they can initiate an immune response or directly damage the skin's barrier, leading to dermatitis.

7.1.2.3 Environmental Factors

Environmental factors can exacerbate dermatitis symptoms. Dry and cold weather can lead to skin dryness and worsen atopic dermatitis, while excessive heat and humidity may trigger sweat-related dermatitis. Exposure to UV radiation from sunlight can also exacerbate some forms of dermatitis.

7.1.2.4 Infections

Bacterial, fungal, or viral infections can lead to a specific type of dermatitis. For instance, fungal infections like ringworm can cause dermatophytid dermatitis, while herpes simplex virus can lead to herpetic dermatitis.

7.1.2.5 Stress and Emotional Factors

Emotional stress and anxiety can exacerbate dermatitis symptoms. Stress can weaken the immune system and trigger inflammatory responses, worsening the condition.

7.1.2.6 Hormonal Changes

Hormonal fluctuations, such as those occurring during pregnancy or menstruation, can influence the severity of dermatitis. Some individuals experience worsening symptoms during these periods.

7.1.2.7 Food Allergies

In some cases, food allergies can contribute to dermatitis, particularly in infants and young children. Common allergenic foods include milk, eggs, peanuts, and soy.

7.1.2.8 Medications

Certain medications, particularly topical or systemic antibiotics, can cause drug-induced dermatitis as an adverse reaction. Other medications and herbal supplements can also lead to skin reactions.

7.1.2.9 Occupational Exposures

Occupational dermatitis is common among individuals who work in jobs that involve exposure to irritants or allergens, such as healthcare workers, hairdressers, and mechanics. Frequent contact with substances like chemicals, latex gloves, or certain metals can lead to occupational dermatitis.

7.1.2.10 Inflammatory Mediators

Dermatitis is fundamentally an inflammatory disorder of the skin. At the molecular level, the condition is characterized by the release of pro-inflammatory mediators. Interleukin-1 (IL-1), Interleukin-6 (IL-6), and Tumor Necrosis Factor-Alpha (TNF-A) are a few examples of these cytokines. The orchestration of the immunological response in the skin is greatly aided by these cytokines. When the skin's barrier is compromised, either due to genetic factors or external irritants, these cytokines are released by immune cells and resident skin cells.

7.1.2.11 Immune Cell Activation

Dermatitis involves the activation of various immune cells. T lymphocytes, specifically Th1 and Th2 subsets, are central to the inflammatory response. Th2 cells, which produce cytokines including interleukin-4 (IL-4) and interleukin-13 (IL-13), are notably linked to atopic dermatitis. These cytokines contribute to inflammation, pruritus (itching), and the characteristic skin changes seen in dermatitis.

7.1.2.12 Epidermal Barrier Dysfunction

Maintaining the integrity of the skin's epidermal barrier is crucial in preventing dermatitis. Molecularly, this involves the role of proteins like filaggrin, which are essential for skin barrier function. Filaggrin genetic abnormalities are linked to a weakened skin barrier, making it easier for irritants, allergenic substances, and pathogens to infiltrate the skin. This breach in the barrier sets the stage for an inflammatory response.

7.1.2.13 Inflammatory Mediators and Vasodilation

Dermatitis is often marked by itching, redness, and swelling, which result from the release of histamine, prostaglandins, and leukotrienes. These molecules mediate vasodilation and increased vascular permeability, leading to the characteristic signs of inflammation observed in dermatitis.

7.1.2.14 IgE-Mediated Allergic Reactions

In certain forms of dermatitis, especially allergic contact dermatitis and atopic dermatitis, In reaction to allergens, the immune system makes a lot of immunoglobulin E (IgE) antibodies. The release of histamine and other inflammatory compounds is triggered when IgE attaches to mast cells. This process amplifies the allergic response and exacerbates dermatitis symptoms.

7.1.2.15 Skin Microbiome Alterations

Recent studies indicate that dermatitis severity and development may be affected by changes in the skin microbiota. Alterations in the skin's microbial community can contribute to inflammation by interacting with the immune system.

7.1.2.16 Neurogenic Inflammation

Dermatitis often involves neurogenic inflammation, where nerve fibers in the skin release neuropeptides that contribute to itching and inflammation. This neurogenic component further intensifies the discomfort associated with dermatitis.

7.1.2.17 Genetic Factors

Genetic factors play a significant role in some forms of dermatitis. Specific gene mutations, such as filaggrin mutations in atopic dermatitis, can make individuals more susceptible to skin barrier dysfunction and inflammation.

7.1.2.18 Therapeutic Potential of Herbal Honey in Managing Dermatitis: A Biomedical Perspective

Herbal honey's therapeutic effects in dermatitis can be attributed to several mechanisms (Fig. 7.5).

7.1.2.19 Antimicrobial Properties

Dermatological conditions are associated with microorganisms, including atopic dermatitis and wound infections brought on by *Staphylococcus aureus*, *Pseudomonas aeruginosa*, and *Escherichia coli*. Pityriasis versicolor, seborrheic dermatitis, atopic dermatitis, and psoriasis are a few of the skin disorders that have been linked to Malassezia yeasts. Traditional treatments have limitations, such as skin thinning with corticosteroids and the potential for skin cancer with UV radiation therapy. In

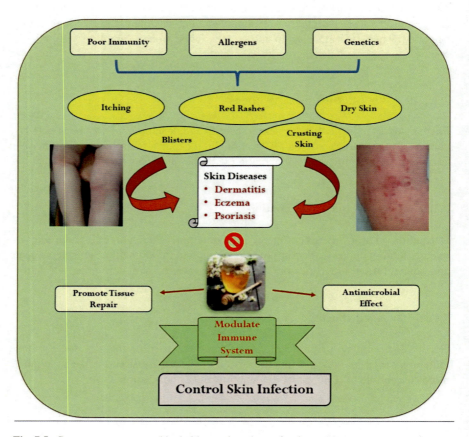

Fig. 7.5 Causes, symptoms, and herbal honey-based cure for dermatitis

contemporary clinical practice, herbal honey has well-documented antimicrobial properties. It contains hydrogen peroxide, low water activity, and high sugar content, all of which contribute to its capacity to prevent the development of bacteria and fungi. In dermatitis cases where bacterial or fungal infections exacerbate the condition, applying herbal honey topically may help reduce microbial load on the skin and support the healing process. Manuka honey is used in a variety of gamma-irradiated gels, ointments, and dressings.

7.1.2.20 Immunomodulating Properties

Despite the fact that the actual causes of some skin ailments, like psoriasis, atopic dermatitis, and contact dermatitis are still not fully known, these conditions are frequently categorized as immune-mediated conditions. The development of these disorders is thought to be significantly influenced by the immune system. They typically react favorably to therapy with immunomodulating drugs such as corticosteroids or UV radiation therapy. It has been demonstrated by recent in vitro experiments that herbal honey may influence the immune system. Researchers have shown, for instance, that herbal honey inhibits the production of specific immune-related proteins and genes in human keratinocyte cells and stimulates the expression of a number of immune-related genes in human primary keratinocytes. However, considering the intricacy of the immune system's function in skin illnesses, it is difficult to predict the specific benefits of honey in treating immune-mediated skin problems without further research.

7.1.2.21 Antioxidant Properties

Herbal honey has an extended history of being used medicinally and gained attention in recent years for its potential therapeutic benefits in managing dermatitis, a group of inflammatory skin disorders. One of the key mechanisms through which honey may alleviate dermatitis symptoms is its antioxidant activity. Dermatitis, including conditions like contact dermatitis and atopic dermatitis, is often characterized by inflammation and oxidative stress in the skin. When the body's capacity to eliminate dangerous reactive oxygen species (ROS) with antioxidants and their synthesis are out of balance, oxidative stress results. The resulting imbalance has the potential to harm tissue and exacerbate dermatitis symptoms. Honey contains a variety of natural antioxidants, including flavonoids, polyphenols, and ascorbic acid (vitamin C). These antioxidants have been shown to scavenge ROS and reduce oxidative stress in the skin. By doing so, honey may help mitigate the inflammatory response and alleviate the symptoms associated with dermatitis.

Numerous research studies have looked into the antioxidant qualities of various kinds of herbal honey. Because of high quantities of methylglyoxal (MGO), herbal honey has exceptionally strong antioxidant action. MGO is known for its strong antioxidant properties and contributes to the overall therapeutic potential of herbal honey in dermatological applications. While herbal honey's antioxidant activity shows promise in managing dermatitis, it is essential to note that not all types of honey are the same. The primary factor influencing honey's antioxidant activity is its botanical origin, while processing, handling, and storage have a minor impact. Herbal honey's antioxidant potential is closely linked to its total phenolic content, with dark honey varieties generally having higher phenolic levels and greater antioxidant capacity. This connection between herbal honey's color and its antioxidant power is due to the diverse floral sources that contribute unique phenolic profiles to the honey.

7.1.2.22 Wound Healing and Tissue Repair

Herbal honey has been recognized and used for its therapeutic properties in traditional medicine across various countries. It is utilized in Malaysian tradition to treat burns, diabetic wounds, furuncles, and carbuncles. In managing eczema, reducing inflammation, and speeding up the healing of wounds, traditional Persian medicine has proven to be successful. Herbal honey is used to heal cuts, wounds, eczema, dermatitis, burns, skin conditions, and Fournier's gangrene in the Indian subcontinent's traditional medical system, Ayurveda. Quranic medicine in Pakistan has recommended a combination of herbal honey and cinnamon powder for pustules, eczema, ringworm, and various skin conditions. These traditional uses of herbal honey remain significant, especially in developing countries where indigenous medicine plays a primary role in healthcare. In modern clinical practice, herbal honey obtained from the honeybees that feed on the manuka flowers in New Zealand (*Leptospermum scoparium*) is utilized topically for managing wound infections. Many nations, including Australia, New Zealand, Europe, the United States of America, Canada, and Hong Kong, have granted clinical approval for this particular variety of honey. Revamil honey, which is made in the Netherlands in partnership with the University of Wageningen and the Academic Medical Centre in Amsterdam, is another medical-grade honey that is frequently used for wound care. Honey possesses beneficial properties when used as a wound treatment. It has been observed to have cleansing properties, promoting tissue regeneration, and reducing inflammation. Moreover, honey-impregnated pads can serve as a non-adhesive dressing for tissues. Clinical studies have shown that using herbal honey dressing has produced better outcomes than using other techniques including amniotic membrane dressing, silver sulfadiazine dressing, and boiling potato peel dressing. Notably, herbal honey dressing has been linked to quicker recovery, a lower risk of contracture, and less scarring.

7.1.2.23 Humectant Properties

Herbal honey is a natural humectant, meaning it can attract and retain moisture. Dermatitis often involves dry, flaky skin, and herbal honey can help moisturize and hydrate the affected areas. Proper skin hydration is crucial for managing dermatitis symptoms.

7.1.2.24 Reducing Itching and Discomfort

Herbal honey's soothing properties can help reduce itching and discomfort associated with dermatitis. When applied topically, it can provide relief from the persistent itching that is characteristic of many dermatitis forms.

Conclusively we can say that herbal honey's potential as a natural remedy for dermatitis is a topic of growing interest in the medical community. Its historical use, rich chemical composition, and various therapeutic mechanisms suggest that it may offer relief to individuals suffering from dermatitis. However, it is important to note that while promising, more rigorous clinical studies are needed to establish its efficacy conclusively. Patients considering honey as a complementary treatment should consult with a healthcare professional to ensure safe and appropriate use. While herbal honey may not serve as a stand-alone solution for dermatitis, its inclusion in dermatological care as a complementary treatment option holds promise and deserves further exploration.

7.1.3 *Herbal Honey and Eye Disorders*

The human eye, also known as the "window to the soul," is our main sensory organ for taking in our environment. The ability to see clearly gives us the ability to connect with people, appreciate the beauty of our environment, and successfully negotiate the challenges of daily life. However, this sensitive and complex sensory system is prone to a number of illnesses and ailments that can damage our vision and disturb its function. Eye problems cover a wide spectrum of illnesses, from small annoyances and pain to serious conditions that could impair eyesight. Genetics, aging-related changes, environmental variables, infections, and systemic disorders are only a few of the many causes. Common eye conditions like conjunctivitis, dry eye syndrome, and styes can greatly interfere with daily activities, leading people to look for relief and treatments.

There has been an increase in interest in alternative and complementary treatments for various eye problems in recent years. The use of herbal treatments, especially herbal honey, is one such strategy. Bees naturally generate honey, which has been praised for its possible medicinal benefits. These benefits include wound healing, antibacterial, anti-inflammatory, and antioxidant capabilities.

Here we are to examine common eye conditions, their effects on human vision and general health, and the possible therapeutic use of herbal honey. We will examine the characteristics of herbal honey, its historical application in eye care, current study findings, and the critical factors to consider and safety measures to take when using this natural medicine. While honey shows potential as a complementary therapy, it should always be used in conjunction with expert medical guidance and with great caution. Eye problems, ranging from slight irritations to serious vision impairments, can drastically lower our quality of life. While conventional medical procedures are frequently advised, some people look into complementary and alternative therapies, such as the use of herbal treatments like honey to treat eye issues. This chapter examines common eye conditions and considers how herbal honey might be used to treat them.

7.1.3.1 Common Eye Disorders

The eyes are frequently referred to as our windows to the world since they are so important to our daily existence. These complex and sensitive sensory organs let us appreciate the beauty of the universe and understand our surroundings clearly. However, a variety of internal and environmental factors can lead to eye disorders that harm our general health and vision. The most prevalent eye conditions that people may experience over their lifetimes are discussed in this section.

7.1.3.2 Conjunctivitis (Pink Eye)

- **Description:** Pink eye, or conjunctivitis, is an inflammation of the conjunctiva, the thin, transparent membrane that lines inside of the eyelids and covers the white area of the eye.
- **Causes:** This condition may be brought on by bacteria, viruses, irritants, or allergies. It is very contagious and spreads quickly, especially in crowded settings.
- **Symptoms:** Redness, itching, tears, discharge, and impaired vision are symptoms. It may come with cold- or flu-like symptoms, depending on the reason.

7.1.3.3 Dry Eye Syndrome

- **Description:** The condition known as dry eye syndrome is brought on by inadequate or poor-quality tear production, which leaves the surface of the eye unnecessarily lubricated.
- **Causes:** Aging, certain drugs, certain environmental factors (such as dry or windy weather), and systemic disorders (such as Sjögren's syndrome) are all contributors to dry eye syndrome.
- **Symptoms:** Patients could feel uncomfortable, hot, itchy, gritty, and have blurry vision. Chronic discomfort and, in severe cases, corneal damage are both consequences of dry eye.

7.1.3.4 Stye (Hordeolum)

- **Description:** A stye, also known as a hordeolum, is a painful, pimple-like protrusion on the eyelid that develops when the oil glands in the eyelids become infected with bacteria.

7.1 Herbal Honey in Communicable Diseases

- **Causes:** The bacteria Staphylococcus aureus frequently cause styes. They could appear if the oil glands in the eyelid get clogged up or inflamed.
- **Symptoms:** Localized erythema, edema, soreness, and pain are common symptoms. The outer or inner eyelid can develop styes.

These typical eye problems can range from minor annoyances to conditions that have a considerable impact on everyday living and eyesight. Some people look for alternative therapies, such as herbal medicines like honey, as complementary possibilities for relief from these disorders, even though they are normally addressed by conventional medical treatments and eye care procedures. The use of herbal honey in treating these eye conditions will be discussed in more detail in the sections that follow, taking into account both conventional wisdom and cutting-edge scientific knowledge. It is crucial to stress that, in order to protect eye health, any investigation into alternative therapies should be performed with caution and under the supervision of medical professionals.

7.1.3.5 Herbal Honey in Cure of Eye Diseases

Herbal honey, which is honey that has been infused with different plant and medicinal herb extracts, may contain qualities that make it effective for eye care. Depending on the particular herbs used in the infusion, these qualities may include antibacterial, anti-inflammatory, antioxidant, and wound healing benefits. For instance, calendula or chamomile-infused honey may have anti-inflammatory effects that can help calm itchy eyes or lessen redness. Herbal honey with flavors of thyme or rosemary may have antibacterial qualities that are helpful in treating eye infections. Understanding the qualities of herbal honey becomes essential when it comes to eye health and the potential treatment of common eye problems.

7.1.3.6 The Properties of Herbal Honey Useful in Eye Care

Herbal honey is a proven eye tonic being used in cleaning of eyes and curing minor infections. The mechanisms and properties of honey in the therapeutics of eyes are as follows (Fig. 7.6).

7.1.3.7 Antibacterial Properties

The antibacterial effects of herbal honey may be helpful in treating some eye conditions. Its low water content and natural sugars make it difficult for germs to grow, and its mild acidity can keep pathogens at bay. A further layer of antibacterial defense is provided by honey's glucose oxidase enzyme, which also produces trace levels of hydrogen peroxide. Because of these properties, herbal honey may be used as a treatment for ailments like conjunctivitis, which can result in bacterial

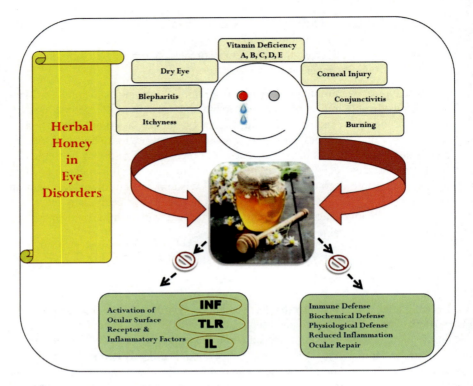

Fig. 7.6 Herbal honey therapeutics in eye disorders

infections of the mucous membranes of the eyes. To ensure purity, sterility, and proper dilution—all important criteria in assuring herbal honey's safety and effectiveness in treating eye disorders—it is imperative to speak with a healthcare practitioner before utilizing herbal honey as an eye treatment.

7.1.3.8 Anti-Inflammatory Effects

The anti-inflammatory properties of herbal honey may have advantages in treating several inflammatory eye conditions. Herbal honey's calming and anti-inflammatory characteristics can be used to treat conditions like conjunctivitis, which cause inflammation and redness of the conjunctiva of the eye. Inflammation can be reduced by herbal honey's natural ingredients, such as flavonoids and phenolic compounds, while its moisturizing properties can relieve dry eyes. However, as purity, sterility, and appropriate application techniques are critical to ensuring safety and efficacy in treating these conditions, it is important to seek the advice of a healthcare professional before utilizing honey as an adjuvant therapy for eye disorders.

7.1.3.9 Antioxidant Activity

The antioxidant activity of herbal honey can be a valuable asset in managing certain eye disorders that involve oxidative stress and damage to ocular tissues. Conditions like age-related macular degeneration (AMD) and cataracts are often associated with the harmful effects of free radicals and oxidative damage. Herbal honey with its natural antioxidants like flavonoids and phenolic compounds can help neutralize these free radicals, potentially mitigating the risk and progression of such disorders. Nevertheless, it is crucial to remember that honey should not be the sole treatment for eye disorders, and patients should always seek professional medical guidance for comprehensive eye care and disease management.

7.1.3.10 Wound Healing

Certain eye conditions, particularly those involving corneal abrasions or epithelial damage to the eye's sensitive surface, may benefit from the wound healing abilities of honey. The antibacterial and anti-inflammatory properties of herbal honey help hasten the recovery of eye wounds while lowering the danger of infection. While its true that honey has long been used for its wound healing properties, it is important to stress that eye injuries or disorders should always be treated by a doctor or ophthalmologist, and that using herbal honey in these situations should always be done so under their direction and supervision to ensure safety and efficacy in the healing process.

7.1.3.11 The Use of Herbal Honey in Eye Care

Traditional Uses

Throughout history, herbal honey has held a revered place in traditional medicine and home remedies. Its application in eye care has been noted across various cultures. While these practices are not a substitute for evidence-based medical treatments, they provide valuable insights into the potential benefits of herbal honey in preserving eye health.

Soothing Eye Irritations

Traditional medicine systems in many cultures have recommended using honey to alleviate eye irritations, including those caused by allergens or minor infections. The honey's anti-inflammatory and antibacterial properties may offer relief from discomfort.

Eyewashes and Compresses

Herbal honey has been used in the preparation of eyewashes and compresses. Diluted honey solutions have been applied topically to soothe irritated eyes. These practices often involve mixing honey with sterile water or saline solution.

Treatment of Eye Infections

Some traditional remedies have employed honey to address minor eye infections. It is believed that honey's natural antibacterial properties can help combat infection and promote healing. However, this approach should be taken cautiously and under professional guidance.

Contemporary Research

While traditional practices provide anecdotal evidence, contemporary research has delved deeper into the potential use of herbal honey in eye care. Scientific studies have explored the mechanisms through which honey may benefit eye health. Although these findings are promising, it is important to note that more research is needed to establish definitive guidelines for the safe and effective use of honey in treating eye disorders.

7.1.3.12 Conclusion

The exploration of herbal honey's potential role in addressing common eye disorders brings to light a fascinating intersection of traditional wisdom and contemporary research. While honey has been celebrated for its antibacterial, anti-inflammatory, antioxidant, and wound healing properties, its application in eye care warrants careful consideration and professional guidance. Here, we delved into the realm of common eye disorders, such as conjunctivitis, dry eye syndrome, and styles, and discussed the historical use of honey in soothing eye irritations and minor infections. We also explored the promising findings from contemporary research regarding herbal honey's antibacterial activity, anti-inflammatory effects, and potential to promote tissue repair and wound healing in ocular tissues.

However, it is crucial to reiterate that herbal honey should not replace evidence-based medical treatments for eye disorders. Instead, it can be viewed as a complementary approach, offering relief from symptoms such as irritation and inflammation. Safety precautions, including proper honey sterilization and dilution, consultation with healthcare providers, and monitoring for adverse effects, are paramount. Individuals with allergies to bee products should exercise extreme caution, and any adverse reactions should be promptly reported to healthcare professionals. Furthermore, herbal honey should always be used under the guidance of qualified experts, particularly ophthalmologists. As we conclude this exploration, we underscore the importance of preserving eye health and seeking professional medical advice for any eye-related issues. While the potential benefits of herbal honey are promising, adherence to established medical treatments and guidelines remains essential. The interplay between traditional remedies and modern ophthalmology continues to provide fertile ground for research and discovery, ensuring that our vision remains a cherished and well-protected sense.

7.1.4 Herbal Honey and Hypersensitivity

Honey has fascinated people for thousands of years because of its golden color and ageless attractiveness. Beyond being a culinary delight, it has established a position in folklore and conventional medicine and is frequently praised for its extraordinary healing power. The fascinating world of herbal honey and its possible uses in the area of hypersensitivity—a condition where the body's immune system frequently assumes the lead—are the subjects of this chapter's investigation. Asthma, atopic dermatitis, allergic rhinitis, and a wide range of allergic reactions are all considered to fall under the umbrella term of "hypersensitivity," which refers to a spectrum of immune-driven responses. Even while traditional medical techniques for treating these disorders have come a long way, research into complementary and alternative therapies is still very active. Its complex chemical makeup is what draws people's interest in honey's potential role in treating hypersensitivity. A complex mixture of sugars, amino acids, vitamins, minerals, enzymes, and phytochemicals make up the sweet nectar known as herbal honey. The solutions to easing the relentless symptoms of hypersensitivity may be found in this diverse collection of natural substances (Fig. 7.7).

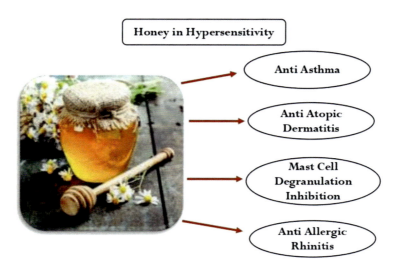

Fig. 7.7 Role of herbal honey in managing hypersensitivity

7.1.4.1 Anti-Asthma Benefits of Herbal Honey

Asthma is a chronic respiratory condition characterized by airway inflammation, constriction, and increased mucus production, resulting in symptoms like wheezing, shortness of breath, and coughing. Although the majority of asthma care involves taking prescribed drugs, complementary and alternative therapies—including the possible contribution of herbal honey—are gaining popularity. The potential anti-asthma benefits of herbal honey have been studied. Herbal honey has a diverse makeup of natural chemicals. Here, we examine the herbal honey's several applications in the treatment of asthma:

7.1.4.2 Anti-Inflammatory Properties

Inflammation of the airways is one of the main processes underlying asthma. Flavonoids and polyphenols, which are naturally occurring anti-inflammatory substances found in herbal honey, may aid in reducing asthma-related inflammation. These substances may function to reduce the activity of inflammatory mediators in the airways, which may lessen the intensity of asthma symptoms.

7.1.4.3 Airway Soothing Effects

When consumed, herbal honey's viscosity and texture can have a calming impact on the airways and throat. Because inflammation and irritation can be uncomfortable during asthma attacks, this relaxing characteristic can be very helpful. Herbal honey can aid in soothing throat soreness and bring about a sensation of relief when swallowed.

7.1.4.4 Antioxidant Defence

Increased oxidative stress in the airways, which results in lung tissue destruction, is linked to asthma. Ascorbic acid and flavonoids, two powerful antioxidants found in honey, can assist in scavenging dangerous free radicals and defending lung tissue. Herbal honey may help to slow the development of asthma by lowering oxidative stress.

7.1.4.5 Cough Suppression

An asthma symptom that is frequently present yet can be quite distressing is persistent coughing. Long-time sufferers of coughing have turned to honey as a natural cure. It may help lessen the frequency and intensity of coughing caused by asthma by coating the throat and having a calming effect.

7.1.4.6 Immune Modulation

An aberrant immunological response in the airways contributes to asthma. Numerous bioactive substances found in honey have the potential to affect the immune system. While further research is required, some studies indicate that herbal honey may have the ability to modify immunological responses, possibly lowering the immune system hyper-reactivity that underlies asthma.

7.1.4.7 Honey as a Potential Treatment for Atopic Dermatitis

Eczema, also referred to as atopic dermatitis, is a chronic inflammatory skin condition marked by red, itchy, and irritated skin. While there are many therapies available, some people use natural solutions like herbal honey to relieve the atopic dermatitis symptoms. It is thought that herbal honey, which has antibacterial and anti-inflammatory qualities, may aid people who have this skin problem. We examine herbal honey's possible utility in treating atopic dermatitis here:

7.1.4.8 Antimicrobial Properties

Herbal honey's inherent antibacterial qualities have long been acknowledged. It includes ingredients like hydrogen peroxide and others that can help fight bacterial and fungal diseases. Applying herbal honey topically may aid in preventing or controlling secondary skin infections that aggravate atopic dermatitis, lessening the severity of symptoms.

7.1.4.9 Anti-Inflammatory Effects

Atopic dermatitis is characterized by inflammation. Flavonoids and polyphenols, two anti-inflammatory substances found in honey, can help lessen swelling, itching, and redness. Herbal honey may have a calming effect when applied to the skin's affected areas, helping to lessen some of the agony brought on by atopic dermatitis.

7.1.4.10 Moisturizing Properties

For people with atopic dermatitis, dry skin is a common issue. Herbal honey is a naturally occurring humectant, which means it can draw in and hold moisture. This aids in moisturizing and hydrating the skin. Herbal honey has the ability to form a barrier that stops moisture loss, which is advantageous for those with dry, eczematous skin.

7.1.4.11 Wound Healing and Skin Repair

Since ancient times, herbal honey has been used to treat wounds, and research has shown that it can encourage tissue repair and regeneration. Herbal honey may enhance the healing process in the context of atopic dermatitis, where the skin barrier is frequently weakened, helping the skin to recover and keep its integrity.

7.1.4.12 Herbal Honey's Potential Role in Mast Cell Degranulation Inhibition

Mast cells are an immune cell subtype that is essential to the body's defence against external intruders. Mast cells, however, can over-activate and prompt disproportionate immunological responses in disorders like allergies and asthma. Mast cells degranulate, releasing a variety of inflammatory compounds, including histamines, which support allergic reactions. It has been researched whether honey, which contains a complex mixture of natural substances, has the ability to prevent mast cell degranulation. Here, we explore the potential role of herbal honey in inhibiting mast cell degranulation:

7.1.4.13 Bioactive Compounds

Bioactive substances such as flavonoids, phenolic acids, and other antioxidants are abundant in herbal honey. The ability to prevent mast cell degranulation has been investigated for several of these substances, such as quercetin. A flavonoid renowned for its anti-inflammatory qualities, and quercetin may assist in lowering mast cell discharge of histamines and other inflammatory chemicals.

7.1.4.14 Anti-Inflammatory Effects

The anti-inflammatory effects of herbal honey may also prevent mast cell degranulation. Herbal honey may help regulate mast cell function and lessen their propensity to release inflammatory mediators by lowering inflammation. Asthma symptoms and milder allergy reactions could be brought on by this restriction.

7.1.4.15 Antioxidant Defence

Antioxidants included in herbal honey have the ability to counteract dangerous free radicals. The body's excessive oxidative stress may make mast cells more active. Herbal honey may aid in preserving mast cell function at a more balanced level by lowering oxidative stress.

7.1.4.16 Immune Modulation

The effects of herbal honey on the immune system are extensive. According to certain research, herbal honey may have immune-modulating properties that could affect mast cell function. Herbal honey's potential as an immune modulator shows promise in the management of hypersensitivity diseases, while additional research is required to pinpoint the precise pathways.

7.1.4.17 Herbal Honey's Potential as a Remedy for Allergic Rhinitis

An allergic response to allergens in the air, such as pollen, dust mites, and pet dander, causes allergic rhinitis, also known as hay fever. Symptoms of the disease include sneezing, runny or stuffy nose, itchy, watery eyes, and throat irritation. To treat allergic rhinitis, some individuals use natural therapies like herbal honey. Due to the idea of local pollen desensitization, it is thought that herbal honey, which is made from the nectar of flowering plants, may provide relief. We look at how honey might help treat allergic rhinitis here:

7.1.4.18 Local Pollen Exposure

The idea behind using honey to treat allergic rhinitis is based on exposure to local pollen. Pollen from the nearby area may be found in some honey made there. The immune system may get gradually desensitized to these allergens by ingesting small amounts of these pollens through honey, which may lessen the severity of allergic reactions over time.

7.1.4.19 Antioxidant Properties

Antioxidants in honey, like flavonoids and polyphenols, can assist the body fight oxidative stress. Histamines and other inflammatory mediators are released, causing an inflammatory reaction that results in allergic rhinitis. Herbal honey's antioxidants might lessen the inflammatory cascade and lessen the intensity of symptoms.

7.1.4.20 Immune Modulation

Research is still being done to determine whether honey has the ability to alter immunological function. Some research implies that honey may be able to affect the immunological response, albeit this is not yet fully understood. Immune modulation may aid in lowering the excessive allergic reaction to pollen and other allergens in the case of allergic rhinitis.

7.1.4.21 Conclusion

In conclusion, herbal honey shows tremendous promise in treating hypersensitivity, providing a holistic, all-natural method of easing allergic symptoms and reactions. Herbal honey is a promising option for managing hypersensitivity due to its natural anti-inflammatory, antibacterial, and antioxidant characteristics as well as the potential for desensitization through the introduction of allergens in minute amounts. Herbal honey's long history as a traditional cure and its safety profile makes it an attractive option for people seeking alternative or complementary ways to combat hypersensitivity, even if research in this field is continuing and the efficacy may vary among persons and certain allergies. It will need additional research and clinical trials to fully understand honey's therapeutic potential in this situation, which will lead to the development of more efficient and comprehensive treatment choices for those dealing with hypersensitivity disorders.

7.1.5 Herbal Honey and Influenza

Herbal honey is prized for its many beneficial qualities and its sweet, golden charm in the world of natural medicines and conventional healing techniques. Herbal honey has become recognized as a potential source of comfort and relief in the setting of respiratory disorders, where discomfort, coughing, and sore throats frequently accompany infections like influenza. The viral infection known as influenza, or "the flu," spreads quickly through an area and causes a variety of symptoms, such as fever, chills, coughing up blood, sore throat, and weariness. There is no cure for influenza, but there are a variety of natural cures and home remedies that work to lessen the symptoms and increase the body's resistance. In the middle of the difficulties of influenza, honey provides a calming hug because of its rich composition of natural components. Herbal honey offers relief from this viral foe that many have sought, from its soothing touch on a sore throat to its potential to soothe persistent coughs. In this investigation, we examine the potential function of herbal honey in the treatment of influenza. We investigate its capabilities as a soothing for sore throats, a cough suppressor, an immune system booster, a hydrator, and a general comforter. Although herbal honey does not specifically combat the influenza virus, it provides comfort from the symptoms that often accompany the infection (Fig. 7.8).

7.1.5.1 Benefits of Herbal Honey in Influenza Management

Herbal honey may aid in the treatment of influenza by easing some flu-related symptoms. While herbal honey cannot treat the flu virus, it can lessen the discomfort brought on by the condition. It is a safe and relaxing natural cure for throat discomfort brought on by the flu because it works especially well at easing sore throats and coughing. The antibacterial qualities of herbal honey may also aid in

7.1 Herbal Honey in Communicable Diseases

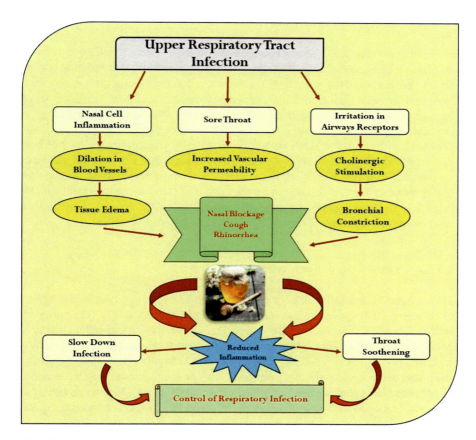

Fig. 7.8 Herbal honey therapeutics in upper respiratory tract infection

general healing, and its consumption helps improve hydration. In extreme circumstances, it is crucial to keep in mind that herbal honey should only be used as a supportive measure and not as a replacement for medical guidance or antiviral medications. However, herbal honey holds promise in providing relief and comfort during an influenza attack as a safe and natural therapy. We examine herbal honey's possible utility in the treatment of influenza here:

7.1.5.2 Soothing a Sore Throat

A sore throat is a typical and frequently upsetting symptom of many illnesses, including viral infections like the flu. It is characterized by discomfort, anguish, and scratchiness. Herbal honey, a beloved natural medicine for its calming effects, can provide comfort and relief for a sore throat. We examine how herbal honey helps to relieve this painful condition here.

7.1.5.3 Coating and Protection

Herbal Honey produces a soothing and protective layer over the inflamed throat and respiratory passages during influenza, which can benefit from its coating and protecting properties. This coating effect aids in reducing the discomfort, inflammation, and irritation that frequently accompany the flu. Additionally, it can act as a barrier, protecting the irritated mucous membranes from additional irritation and outside allergens. Herbal honey acts as a barrier to keep germs out, making it easier for those with the flu to swallow and breathe easily. This promotes overall health and speeds up the healing process. Although herbal honey can be a helpful natural therapy for symptom management, particularly when it comes to easing the throat and respiratory tract, it should not be regarded as a stand-alone treatment for influenza and medical guidance should be sought, when necessary, particularly for severe cases of the illness.

7.1.5.4 Anti-Inflammatory Action

Herbal honey's anti-inflammatory properties can be especially helpful during influenza when the body's immune reaction frequently causes severe respiratory tract inflammation. The natural benefits of herbal honey, such as its anti-inflammatory and antioxidant components, might lessen the discomfort and irritation brought on by the inflammation. It relieves soreness and discomfort brought on by the flu by soothing the irritated throat and respiratory passages. This anti-inflammatory action is particularly beneficial because it aids in both comfort and the body's healing process, making it easier for people with influenza to breathe and swallow, which promotes quicker recovery and general well-being throughout this respiratory infection. But it is important to keep in mind that while herbal honey might help control symptoms, it shouldn't take the place of expert medical care, especially in severe cases of influenza.

7.1.5.5 Antimicrobial Benefits

When a sore throat is brought on by a bacterial illness, herbal honey's antibacterial qualities are beneficial. Inhibiting bacterial growth in the throat using honey could hasten the healing process. It is crucial to remember that many sore throats are brought on by viruses, such as the influenza virus, in which case the soothing effects of herbal honey are more pertinent.

7.1.5.6 Cough Suppression

Herbal honey is an effective cough suppressant, especially during influenza, since it provides a natural treatment for the cough that frequently accompanies the sickness. Because of its thick texture, it coats the irritated throat and eases the discomfort that causes coughing. Natural sugars in honey act as demulcents, calming the throat and lessening the need to cough. Additionally, herbal honey's antimicrobial qualities can aid in treating the cough's underlying causes, such as bacterial or viral illnesses. Due to its safety and accessibility, herbal honey is a favored and successful alternative to over-the-counter cough drugs, which may have more side effects or be less suitable for children, for treating coughing caused by the flu.

7.1.5.7 Immune Support

Herbal honey includes antioxidants and other bioactive substances that can support the immune system, even if they are not direct sore throat therapy. The underlying cause of the sore throat, such as the flu virus, may be addressed by the body's natural healing processes with the help of this assistance. A sore throat, a typical influenza symptom, can be made better by the thick, viscous nature of herbal honey. It coats the throat and can lessen irritation and discomfort, which eases the discomfort of swallowing.

7.1.5.8 Hydration and Comfort

Influenza puts a significant amount of stress on the body and frequently results in dehydration because it is characterized by fever, sweating, and a general feeling of malaise. It is crucial to be properly hydrated at this period for a number of reasons. Dehydration can make flu symptoms worse, delay recovery, and make the immune system less effective at battling viral illness. Additionally, it may result in a dry mouth and throat, adding to general discomfort. Herbal honey proves to be a useful ally in the fight against dehydration caused by the virus. When the desire to drink is low, its relaxing and appealing natural qualities increase fluid intake. A plain glass of liquid can be made into a soothing, tasty beverage by adding honey to warm water or herbal tea. The act of hydrating is now more appealing and delightful thanks to this small improvement. The delicious flavor of herbal honey serves as a motivator, especially when loss of appetite is a typical influenza symptom. Herbal honey aids in sustaining the ideal fluid balance in the body by making hydration attractive. As a result, it helps to lessen the severity of flu symptoms, encourages a quicker recovery, and improves general comfort while battling this formidable viral foe.

7.1.5.9 Restful Sleep

The body's natural healing and recovery processes depend heavily on restful sleep, which is even more important during an influenza infection. The symptoms of influenza, which frequently include fever, coughing, body pains, and congestion, might make it difficult to fall asleep without disturbance. Herbal honey might be of great help in this situation due to its calming qualities. There are a number of reasons why honey can aid in comfortable sleep. First, because of its thick and viscous structure, it can coat and calm the sore mucous membranes of the throat and airways, reducing the discomfort brought on by chronic coughing and throat irritation. This calming impact reduces nighttime disturbances, enabling more uninterrupted and restorative sleep.

7.1.5.10 Antioxidant and Anti-Inflammatory Properties

Herbal honey's natural antioxidant and anti-inflammatory capabilities stand out as crucial elements in easing discomfort and resolving the variety of symptoms that usually accompany this viral infection when it comes to managing influenza. The antioxidants in honey are first and foremost effective inhibitors of oxidative stress, a condition of elevated cellular damage brought on by the body's immunological reaction to the flu virus. This oxidative stress can aggravate symptoms, prolong illness, and cause overall malaise. The antioxidants in herbal honey combat dangerous free radicals, lowering oxidative stress and so minimizing these negative consequences.

The anti-inflammatory effects of herbal honey are also very important in reducing discomfort. Inflammation throughout the body, especially the respiratory system, which is a hallmark of influenza can cause symptoms like headaches, aches in the muscles, and general malaise. Due to herbal honey's anti-inflammatory properties, these symptoms are alleviated and a higher sense of well-being is made possible.

Despite these possible advantages, it is crucial to remember that herbal honey doesn't directly combat the influenza virus. It alleviates symptoms while assisting the body's own healing mechanisms. Since influenza is a viral infection, serious cases may call for the use of antiviral drugs that a doctor has recommended. Additionally, because baby botulism is an uncommon risk, care should be taken when providing honey to children under the age of one.

7.1.5.11 Conclusion

In the world of influenza management, herbal honey proves to be a helpful friend. It does not directly fight the flu virus, but it offers comfort and support during the battle. Herbal honey encourages us to stay hydrated, which is essential for our body to fight the flu and feel better. When we are sick, drinking can be tough, but honey makes it easier and more enjoyable. Herbal honey also helps us sleep better. When you have the flu, you might cough a lot and feel uncomfortable. Herbal honey soothes your throat, reduces the need to cough, and lets you sleep more peacefully. Moreover, herbal honey's antioxidants and anti-inflammatory properties are like bodyguards. They protect your body from harmful things and reduce discomfort. So, those pesky headaches and muscle aches caused by the flu become less bothersome. In the end, honey doesn't replace medical treatment for the flu, but its a comforting companion on your journey to recovery. It is like a warm hug, combining tradition with modern ways to help you feel better during the flu.

7.1.6 *Herbal Honey and Periodontal Disease*

Periodontal disease is a widespread and significant bacterial infection impacting the gums and the surrounding tissues that provide support to the teeth. It generally progresses through various stages, starting from initial stage gingivitis to progression stage periodontitis, and can lead to tooth loss if left untreated. Globally, periodontal disease is a widespread concern, affecting a significant percentage of people all over the world. According to different international agencies like WHO and the Centers for Disease Control and Prevention (CDC), these diseases are believed to impact approximately 19% of adult population, constituting more than 1 billion cases all over the globe. Prevalence rates vary among different age-groups, with older individuals demonstrating a higher susceptibility to periodontal diseases (Nazir et al. 2020).

Periodontal disease (PD) is mainly triggered by the growth of plaque, a sort of gummy film of bacteria, on gums and teeth. Common symptoms encompass inflamed or red gums, bleeding during brushing or flossing, persistent bad breath, the presence of pus between teeth and gums, alternations in bite or tooth alignment, lose or shifting teeth (Genco and Borgnakke 2013). The main risk factors for periodontal disease are bad oral health, chewing tobacco, salts of toothpaste, etc. Certain medical conditions such as diabetes, pregnancy, age, hereditary or genetic diseases, may also increase the risk of developing periodontal disease (Chen et al. 2022).

Research studies indicate that honey may have potential beneficial effects on periodontal disease. A study demonstrated that manuka honey exhibited antibacterial properties against plaque-associated bacteria, particularly *Streptococcus mutans* (Safii et al. 2017). The experiment involved immersing hydroxyapatite globules in the solution of honey, revealing pH-dependent calcium dissolution. When inoculated with *S. mutans*, both manuka and white clover honey promoted further demineralization (Fig. 7.9).

Herbal honey can also potentially be used in the management of oral bacterial infections. A study showed the effectiveness of honey in thwarting dental caries and gingivitis in orthodontic patients (Atwa et al. 2014). This study revealed that chewing honey resulted in a significant increase in plaque pH compared to sorbitol. Although both honey and sucrose groups initially experienced a pH drop at 5 min, the honey group recovered quickly, thus maintaining a pH level above the critical decalcification threshold. Significantly, bacterial counts, including *Streptococcus mutans*, *Lactobacilli*, and *Porphyromonas gingivalis*, were significantly reduced (Fig. 7.9). Notably, honey exhibited potent antibacterial activity against these strains.

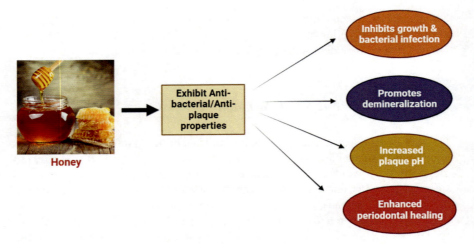

Fig. 7.9 The impact of herbal honey on periodontal disease

The study concludes that honey has the potential to act as an alternative in controlling dental issues and gingivitis, highlighting its promising role in maintaining oral health (Atwa et al. 2014).

Moreover, honey also seems to possess anti-plaque properties, as one of the studies revealed that mouthwash containing multifloral processed honey inhibits the progress of six predominant oral bacterial species, leading to a reduction in plaque formation (Aparna et al. 2012). An additional study unveiled the positive impact of Tualang honey on periodontal therapeutic (Ibrahim et al. 2021) (Fig. 7.9).

7.1.7 Herbal Honey and Herpes Disease

Herpes, caused by the Herpes Simplex Virus (HSV), is a widespread infection known to induce painful blisters or ulcers. This ailment is primarily attributed to HSV-1 and HSV-2. According to a study, approximately 491 million individuals were found to be living with HSV-2 infections, while 358.5 million people had oral HSV-1 infection (James et al. 2020). Additionally, 122–192 million individuals were assessed to have genital HSV-1 infection among those up to 49 years of age. HSV-2 occurrence increases with the passage of age, with rates of 5.4%, 8.3%, and 14.4% among women aged 15–20 years, 21–25 years, and over 25 years of age, respectively (Madhivanan et al. 2009).

HSV-1, generally related to oral or cold sores, is widespread, with a majority of adults being infected by it. Whereas, HSV-2 is associated with genital herpes only. The virus spreads when a healthy person comes in direct contact with disease-ridden or bodily fluids (Zhu and Viejo-Borbolla 2021). The majority of individuals with herpes experience either no symptoms or only mild ones. These symptoms include

7.1 Herbal Honey in Communicable Diseases

painful sores or blisters on the skin or mucous membranes, itching or tingling before the sores appear, flu-like indications or enlarged and tender lymph nodes.

Indeed, there is promising research suggesting the effectiveness of honey in treating herpes infections. Due to the antiviral properties of herbal honey, it may be a possible therapeutic choice for herpes labialis (Fig. 7.10). Studies have indicated that honey can reduce the symptoms and signs of herpetic lesions by inhibiting prostaglandin at the lesion site (Asghari et al. 2018). Similarly, another study highlights the efficacy of topical honey application in the management of recurrent herpetic lesions, and genital herpes, as compared to acyclovir cream (Al-Waili 2004a, b). The findings indicate that honey treatment for labial herpes and genital herpes resulted in an improvement in attack duration, pain, reduction in crusting occurrence, and a decrease in mean healing time compared to acyclovir (Fig. 7.10). Notably, cases of complete remission were observed with honey.

Moreover, the use of medical-grade honey (MGH) presents a promising alternative for the cure of cold sores. A study revealed a decrease in pain levels and itching among patients with recurrent cold sores, thus potentially offering a dual benefit of increased antiviral activity and enhanced wound healing (Naik et al. 2021; Fig. 7.10). This underscores the potential efficacy of MGH in addressing the symptoms associated with cold sores. Another study demonstrated the usefulness of New Zealand medical-grade kanuka honey when used in comparison with topical acyclovir, for treating herpes simplex labialis, also known as cold sores (Semprini et al. 2019). No significant difference was observed between the two treatments. The primary

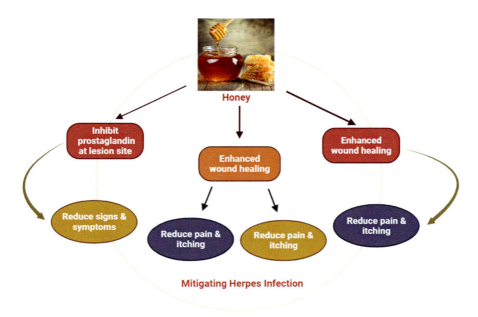

Fig. 7.10 The effect of honey in mitigating herpes infection

outcome, which measured the time to return to normal skin, was comparable for both acyclovir and honey. The study concludes that there was no discernible distinction in effectiveness between the application of topical medical-grade kanuka honey and acyclovir.

7.1.8 Herbal Honey and Viral Hepatitis

Hepatitis is tenderness of the liver. It can be instigated by many factors like viruses, drugs, and toxins. The inflammation or damage to the liver can affect its working leading to decreased processing of nutrients, accumulation of toxins in blood, and decreased infection resistance. The primary reason for hepatitis is viral infection known as viral hepatitis. The hepatitis cursing viruses are Hepatitis A, B, C, D, and E, which infiltrate the liver cells. The viral hepatitis in the early stages is asymptomatic. The viral hepatitis HBV and HCV collectively caused about 1.1 million deaths all around the world in 2019. It has been estimated that about 354 million people are suffering from hepatitis B and C infections (CDC 2019; WHO). World Health Assembly in 2016 set a global target to eradicate hepatitis from a global threat level by the year 2030. They set a goal of a reduction in infections by 90% and a 65% reduction in mortality (Vogt et al. 2022). Viral hepatitis continues to be a hazard to millions of people worldwide, despite advances in medical research, and presents difficulties for both prevention and treatment.

The symptoms of acute viral hepatitis can take weeks to months to manifest and may include symptoms such as fever, no appetite, fatigue, nausea, pain in the abdomen/joints, yellow-colored urine, jaundice, etc. (Banker 2003). Untreated viral hepatitis in severe cases may lead to chronic hepatic inflammation, liver cirrhosis, and even hepatocellular carcinoma. In the case of chronic viral hepatitis, the symptoms sometimes take decades to manifest.

7.1.8.1 Mode of Transmission

Hepatitis A and E viruses spread enterically, yet HAV has more clearly defined modes of transmission than HEV. HEV is zoonotic with unclear mechanisms of transmission, whereas HAV is mainly transmitted by the fecal-oral route. Since vaccinations for infants were advised, the number of new cases of HAV infection has sharply declined in developed nations, and HEV infections are rarely detected. However, these viruses constitute a serious hazard in developing and undeveloped nations with inadequate sanitation and hygiene practices because they spread by fecal-oral pathways, primarily through polluted water supplies or the ingestion of raw food that has been washed in contaminated water (Hofmeister et al. 2019). Human-restricted and waterborne, HEV genotypes 1 and 2 are found in underdeveloped nations, whereas zoonotic genotypes 3 and 4, found in industrialized nations, preferentially infect humans by infected meat intake (Hofmeister et al. 2019). Infected blood and bodily fluid interactions can spread Hepatitis B, C, and D. Shared

7.1 Herbal Honey in Communicable Diseases

needles, unprotected intercourse, and contact with contaminated blood or bodily fluids are among the ways they can spread. Compared to HCV, HBV is more frequently transmitted from mother to child. Since HDV can only infect hosts who are already infected with HBV, it has a more selective host selection process. It is incapable of infecting healthy hosts (Banker 2003; Foster et al. 1998; Gunson et al. 2003). To stop the spread of each hepatitis virus, it is essential to comprehend these routes of transmission and design specific interventions and preventive measures. Education, immunization campaigns, and public health activities are essential for reducing the global effect of viral hepatitis.

Honey is the fermented product prepared by the honeybees from the nectar. It is well-established fact that honey has antibacterial, anti-inflammatory, and immunomodulatory properties. In some cases, the use of honey has been suggested as an antiviral (Kumar et al. 2024). A study by Abdulrhman et al. (2016) showed the potential use of honey in the cure of HAV in children, where 92% of children in the honey group exhibited complete recovery compared to 72% in the placebo group. Honey along with the recovery from HAV also relieved from symptoms such as appetite loss, nausea, fever, and abdominal pain. As a result, the researchers recommended honey as a potential nutritional supplement for children suffering from hepatitis A. Similar results were also presented in a cohort study on HCV treatment (Kotb et al. 2014), where honey, a healthy diet, regular exercise, and abstinence from drugs showed 35% more recovered patients from HCV. These studies have shown the emergence of honey as a promising natural remedy for managing symptoms and promoting recovery in individuals with viral hepatitis (Fig. 7.11).

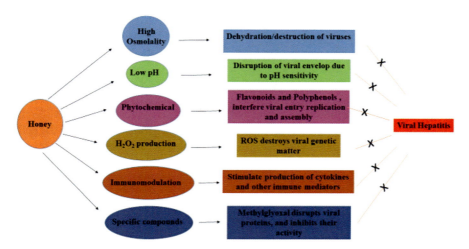

Fig. 7.11 The antiviral mechanism of action of honey. The osmolality of honey is high so it can draw water from the viruses leading to dehydration and destruction of viruses. Secondly, the low pH of honey can disrupt the pH-sensitive envelopes of viruses. Phytochemicals such as flavonoids and polyphenols prevent the entry, replication, and assembly of viral components in the cells. The H_2O_2 production by honey in the body can raise the reactive oxygen species destroying the viral genome. Honey can stimulate the production of cytokines, and it also contains methylglyoxal, where these can lead to immune modulation and disruption of viral proteins, respectively, so that the viruses can be eliminated

7.1.9 Herbal Honey and Gingivostomatitis

It is a medical condition in which oral infection can result in canker sores in the mouth and lips blisters. Stomatitis refers to inflammation in oral tissues including the gum, tongue, lips, inner sides' floor, and roof of the mouth. This disease is most commonly caused by a bacterial or viral infection, usually by herpes simplex virus, which is found in two strains Herpes Simplex Virus Type 1 and 2, both belong to the Herpes virus family, Herpesviridae (Arduino and Porter 2008). It causes herpetic gingivostomatitis, which is characterized by painful oral lesions and a high-grade fever (Aslanova et al. 2023).

HSV-1 consists of linear double-stranded DNA. It is commonly transmitted when a person comes in direct contact with some or another kind of body fluids. HSV-1 contamination is more common and started in early childhood from ages 6 months to 5 years, it may also occur in adults. Children with primary HSV-1 infections might either show no symptoms at all or develop mucocutaneous vesicular eruptions after an approximate one-week incubation period. After a patient contracts the herpes simplex virus, the illness may return to herpes labialis, with sporadic reactivations happening all throughout the patient's life (Arduino and Porter 2008). Most commonly, ocular, face, and oral infections are caused by HSV-1 because of its affinity for the mouth epithelium. Although HSV-1 infection is linked to the majority of instances of herpetic gingivostomatitis, HSV-2 is considered an infection in below the waist (genital region) (Arduino and Porter 2008).

The disease is characterized by a feverish prodrome that is followed by the emergence of painful ulcerative gingiva and mucosal lesions, as well as yellow, perioral, and vesicular lesions. Fever, chills, nausea, loss of appetite foul breath, and anorexia are associated symptoms.

HSV-1 and HSV-2 possess three biological characteristics viz., reactivation, latency, and neurovirulence that have insignificant role in causing illness. These consist of reactivation, latency, and neurovirulence. The capacity to enter and spread throughout the nervous system is known as neurovirulence, and being able to keep a latent infection within a nerve cell is known as latency. Reactivation is the tendency of a disease process to reproduce and start up again after it has been triggered by particular stimuli. Using the same molecular mechanisms, HSV-1 produces herpes gingivostomatitis and ultimately herpes labialis. The herpes simplex virus replicates itself during the pathogenesis of herpes gingivostomatitis, causing cell lysis and ultimately mucosal tissue loss. When abraded surfaces are exposed to HSV-1, the virus can enter and rapidly multiply in dermal and epidermal cells (Aslanova et al. 2023).

Herbal honey has been used therapeutically for a very long time, against various bacterial, microbial, and viral. Gingivostomatitis is a viral disease, highly contagious and spread with direct contact. Honey's natural properties like high osmolality and low pH, as well as its physiologically independent bioactive components, which differ between honey varieties and may have antiviral, anti-inflammatory, and wound healing effects, might be the basis for its medicinal benefits (Semprini

7.1 Herbal Honey in Communicable Diseases

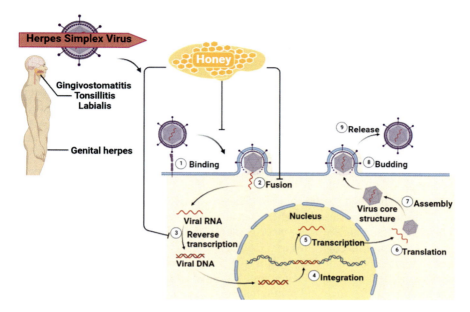

Fig. 7.12 Showing herbal honey therapeutics role in inhibition of viral infection

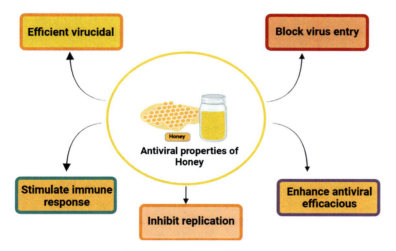

Fig. 7.13 Showing potential antiviral properties of herbal honey

et al. 2019; Figs. 7.12 and 7.13). An abundant amount of phenolics and flavonoid components in honey exhibit antiviral properties. Viral infections lead to disruptions of cell membranes and integration of genome in host genetic machinery, and replicate its own viral genome, leading to lesions and inflammation. Honey and its constituents have cell-arresting, anti-inflammatory, and antioxidant properties.

Recent studies showed herbal honey's antiviral properties in the treatment of gingivostomatitis. Awad and Hamad (2018) assessed the effectiveness of honey, when combined with oral acyclovir, for treating Herpes Simplex Gingivostomatitis (HSGS) in objects aged 2–8 years. The randomized double-blind placebo-controlled trial included 100 participants, comparing the outcomes of this combined therapy against the administration of acyclovir alone. The results demonstrated compelling benefits associated with the use of herbal honey. Children in the study group exhibited a markedly faster resolution of herpetic disorders in comparison to the control group. Moreover, the study group described lower pain, improved eating and drinking capability, and fewer doses of analgesics at various assessment points. Although both groups experienced fever resolution, the study highlighted herbal honey's potential to enhance the therapeutic effects of acyclovir, presenting a promising adjunctive approach for the cure of HSGS in young children.

The study investigated by Schieber et al. (2019) used Kanuka honey against acyclovir for the treatment of herpes simplex labialis (HSL) in 952 participants by using a randomized controlled trial method. Comparing the trial, objects were given either topical honey or 5% acyclovir. No significant differences were observed among the treatments. Both therapies have equal effectiveness for all variables outcomes like healing time and pain relief in between two groups. In the treatment of HSL, herbal honey might be used as an equivalent beneficial option to acyclovir cream, especially in emerging situations like drug resistance and patient groups of lactation and pregnancy, in order to prevent adverse effects.

Münstedt et al.'s (2019) study investigated a systematic review of efficacious of bee products (honey, propolis) in the treatment of symptoms caused by the herpes virus. The studies included 3 trials for honey including 16 patients (8 labial + 8 genital). It was found that herbal honey was highly effective in the treatment of lesions.

Another study reported by Coppola et al. (2023) delves into the clinical management of primary herpetic gingivostomatitis (PHGS), primarily caused by Herpes Simplex Virus 1 (HSV-1). With a global prevalence and the potential for complications, the systematic review identified various treatment regimens for PHGS, including acyclovir, acyclovir combined with honey, fluids, analgesics, and several other symptomatic drugs. Notably, the study highlighted a lack of consensus in therapeutic management, exposing gaps in existing evidence. However, among the treatment strategies, the combination of acyclovir and honey stood out, signifying a probable role for honey in addressing herpetic gingivostomatitis. The challenges in conducting randomized clinical trials were acknowledged, given the rapid onset and remission of the disease, coupled with an investigative delay that diminishes the effectiveness of antiviral drugs. The findings underscore the need for further research and standardization in the clinical management of PHGS, with honey emerging as a promising adjunctive measure warranting exploration.

7.1.10 Herbal Honey and Anal disorders

Anal disorders are known as a group of diseases which impact the anus and rectum. Hemorrhoids, anal fissures, anal abscesses, anal fistulas, and fecal incontinence are common anal and rectal disorders. Depending on the specific disease, anal illnesses may exhibit a variety of symptoms, such as anal discomfort, itching, bleeding, discharge, and trouble passing feces.

Hemorrhoid is a medical condition in which swollen and inflammatory veins are around the anal and lower region of rectum. It resulted in frequent bleeding as well as little discomfort or irritation. Occasionally, a clot or thrombus develops from blood inside a hemorrhoids. It caused excessive strains during bowel movement. It involves the disruption supporting tissue of anal region. The study demonstrated the activation of matrix metalloproteinase (MMP) in hemorrhoids. Level of MMP9 extensively expressed in hemorrhoids which excessively disrupt the extracellular proteins such as fibronectin, elastin, and collagen (Han et al. 2005; Abdellah and Abderrahim 2014).

An anal fissure is a condition in which opening region of anus is break or rupture, that reveal part of the anal canal's muscle fibers. The most frequent reason for an anal fissure is passing of extremely solid or liquid feces. In this condition, passage of feces through anal canal is very difficult (Abdellah and Abderrahim 2014).

7.1.10.1 Anal Abscesses and fistula

An anal gland infection causes anal fistulas and abscesses that can be characterized as inter, isc, peri, or supra based on the position of the abscess (Rickard 2005). Anorectal fistulas and abscesses can cause discomfort, swelling, drainage, bleeding, constipation, and an overall sensation of illness (Abdellah and Abderrahim 2014).

Symptoms of anal disorders (Akkoca et al. 2022; Bharucha and Wald 2010; Mao et al. 2017) are as follows:

- Itching or irritation around the anus
- Pain or discomfort during bowel movements
- Bleeding from the rectum
- Discharge or leakage
- Swelling or lumps near the anus

Honey is composed of a diverse range of additional ingredients and trace concentrations that have a variety of nourishing and biotic impacts. Herbal honey has its ability to serve as a defense mechanism due to containing phenolic compounds (Kurek-Górecka et al. 2020; Fig. 7.14).

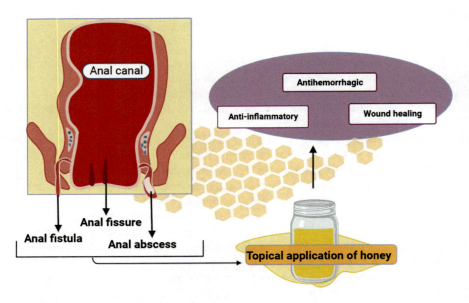

Fig. 7.14 Showing the efficacious role of herbal honey in the treatment of anal disorders

Recent studies showed that honey is highly efficacious in the treatment of anal disorders. It is used topically in the form of a composition mixture or in the form of lotion.

The study by Al-Waili et al. (2006) underscores the potential effectiveness of a topical mixture comprising herbal honey in treating anal fissures and hemorrhoids. A prospective pilot study involving 15 patients demonstrated positive therapeutic effects, with a 12-hour application of the natural formulations significantly reducing bleeding and relieving itching in patients with hemorrhoids. Those with anal fissures experienced substantial reductions in pain, bleeding, and itching after treatment. The scoring method used to assess symptoms indicated significant improvements. Importantly, no bad impact was reported, highlighting the safety of the honey-based mixture. The study concludes that this combination is both safe and clinically effective in treating hemorrhoids and anal fissures, suggesting its potential as a therapeutic option. The findings encourage further exploration through randomized double-blind studies to validate and refine these promising results.

The study by Vlcekova et al. (2012) highlights the potential therapeutic effectiveness of honey in the cure of perianal fistula, a condition commonly associated with inflammatory bowel disease. Over 6 months, treatment with local sterilized honeydew honey resulted in the complete healing and closure of most fistulas in the gluteofemoral region. Notably, honey demonstrated additional benefits by reducing inflammation, pain, and induration in the affected region, positively impacting the patient's mental well-being and overall quality of life. The honey used exhibited potent antibacterial activity, particularly against multi-drug-resistant clinical isolates. The study emphasizes the efficacious application of honey in the treatment of perianal fistula, suggesting its potential as a therapeutic option in cases where conventional methods have failed. The authors advocate for the use of γ-irradiated honey for medical purposes to ensure sterility in the treatment of deep wounds or fistulae. This case underscores the promising outcomes and encourages further exploration of honey's medicinal potential in various clinical contexts.

The study carried out by Razdar et al. (2023) formulates lotion by using two natural flavonoids found in olive oil and honey propolis and evaluates their therapeutic role and adverse effects to those of an anti-hemorrhoid ointment (including hydrocortisone and lidocaine). In the study, a total of 86 participants were in a randomized clinical trial. It consists of case group which is treated with flavonoid lotion and control groups. The study revealed that hemorrhoidal symptoms and complications may be effectively treated with flavonoid lotion. Additionally, compared to regular anti-hemorrhoid ointment, it works faster. Therefore, this flavonoid lotion may be presented as a suitable, accessible, affordable, and convincing therapy candidate of hemorrhoids.

7.2 Herbal Honey in Non-Communicable Diseases

7.2.1 Herbal Honey and Diabetes

Diabetes mellitus is a metabolic disorder which has been known since antiquity. Descriptions of diabetes have been found in the Egyptian, ancient Indian, Greek, as well as Chinese literature. Aretaeus of Cappadocia in second century AD gave the first detailed account of the disease and coined the term diabetes. However, the term mellitus was added to the disease in seventeenth century AD by Thomas Willis. In 1921, Frederick Banting and Charles Best made great contribution in the treatment of diabetes. They isolated insulin from pancreatic islets and administered to patients suffering from type 1 diabetes.

Symptoms of diabetes include frequent urination, high blood sugar level, extreme thirst, nausea and vomiting, blurry vision, more hunger, weakness, loss of weight, etc. Two major forms of diabetes are Type I diabetes or insulin dependent diabetes mellitus (IDDM) and Type II diabetes or non-insulin dependent diabetes (NIDDM). Type I diabetes is generally characterized by the sudden appearance of severe symptoms, deficiency of insulin, and susceptibility to ketosis. Due to destruction or inactivity of the beta cells, pancreas stop producing insulin. In developing countries, IDDM is the most prevalent among children and young adults. Type II diabetes (NIDDM) is a metabolic disorder due to insulin resistance which ultimately leads to rise in blood glucose. Fat, muscle, and liver cells fail to respond to insulin so that blood sugar cannot enter these cells. Chronic hyperglycemia is associated with dysfunction and failure of different organs resulting in increasing levels of morbidity and mortality. Type II diabetes accounts for about 90% of the diabetic population and is the more common form of diabetes.

The prevalence of diabetes mellitus has been increasing day by day. It has been estimated that there were 366 million people in 2011 suffering from diabetes mellitus. It is predicted that this number will increase to 552 million by the year 2030 (Olokoba et al. 2012).

7.2.1.1 Causes of Diabetes

Diabetes mellitus is a syndrome caused either due to deficiency of insulin or impaired effectiveness of insulin's action or combination of these.

Type 1 diabetes occurs due to lack of insulin which is caused by the destruction of insulin-producing beta cells in the pancreas. It is an autoimmune disease in which the immune system of body attacks and damages the beta cells. Destruction of beta cells may take several years, but symptoms of disease generally appear over a short period of time. Heredity plays a significant role in the development of type 1 diabetes. Genes which are transferred from parents to offsprings carry instructions for the synthesis of proteins required for the functioning of body's cells. Some of these genes may determine susceptibility to type 1 diabetes. Some genes carrying instructions for the synthesis of proteins on white blood cells known as human leukocyte antigens (HLAs) are associated with the risk of developing type 1 diabetes. In addition to the HLA genes which are the major risk genes for type 1 diabetes, several other risk genes have also been identified. White blood cells called T cells attack beta cells. The immune systems of the person susceptible to type 1 diabetes counter the insulin considering it as a foreign substance or antigen. As a response, the body produces proteins called antibodies to combat these antigens.

Type 2 diabetes develops when the body can no longer produce enough insulin to compensate for the impaired ability of the cells to use insulin. Genetic factors play an important role in susceptibility to type 2 diabetes. Some genes or their combinations may enhance or reduce the risk of developing the disease. Recently, several gene variants have been identified which influence the susceptibility to type 2 diabetes. It has been found that gene variant TCF7L2 increases susceptibility to type 2 diabetes. Genes may also enhance the tendency of a person to become obese or overweight which is strongly associated with the development of type 2 diabetes. As a consequence, muscle, fat, and liver cells couldn't respond to insulin, compelling the pancreas to produce extra insulin. As the insulin production is not sufficient due to disfunction of beta cells, there is rise in glucose level, causing diabetes.

7.2.1.2 Management of Diabetes

It is a well-known fact that there is no cure for diabetes. However, treatment of diabetes may involve the following measures:

1. Constant monitoring of blood sugar
2. Oral medications to lower the blood glucose

3. Injections of insulin
4. Weight control
5. Healthy diet
6. Regular exercise

7.2.1.3 Therapeutic Potential of Herbal Honey in Managing Diabetes

In spite of the availability of numerous antidiabetic drugs, diabetes mellitus still remains a challenge worldwide. Unfortunately, there is no cure for diabetes; however, the risk of long-term complications can be reduced by controlling blood sugar levels through regular exercise, healthy diet, and proper medication. Herbal treatment of diabetes through traditional medicines has shown more promising results to reduce the ill effects of the disease. Medicinal plants which are generally used for the treatment of diabetes include *Acacia nilotica, Aegle marmelos, Allium sativum, Aloe vera, Annona squamosa, Azadirachta indica, Berberis vulgaris, Catharanthus roseus, Cinnamomum verum, Curcuma longa, Eugenia jambolana, Gymnema sylvestre, Momordica charantia, Murraya koenigii, Ocimum sanctum, Phyllanthus emblica, Phyllanthus niruri, Picrorhiza kurroa, Pterocarpus marsupium, Swertia chirayita, Tinospora cordifolia, Trigonella foenum-graecum, Withania somnifera, Zingiber officinale,* etc. It has been demonstrated experimentally that several herbal extracts regulate the metabolic processes like glycolysis, gluconeogenesis, Krebs cycle, synthesis and release of insulin, synthesis of glycogen, cholesterol synthesis, and metabolism and absorption of carbohydrates.

The exact mechanism of the antidiabetic effect of honey is complex. It has been found that fructose in honey modulates its hypoglycemic or antidiabetic effect. Based on the earlier works, it has been suggested that the fructose content of honey negatively correlates with glycemic index. Small fractions of fructose reduce glucose level in blood through increased hepatic glucose uptake by activating glucokinase. Oligosaccharides present in honey have also been reported to contribute to the antidiabetic effect of honey through modulation of gut microbiota or through their systemic effects. Fructose and other oligosaccharides present in honey delay digestion and intestinal absorption of glucose which results in reduced glycemia. Moreover, some of the minerals present in honey like chromium have been reported to cause reduction in blood glucose and regulation of the secretion of insulin from β cells of pancreas. Additionally, zinc and copper present in honey also contribute in the reduction of glucose level in blood. Honey causes increased secretion of insulin and modulation of appetite regulating hormones which ultimately leads to improved glycemic control.

It has been found that honey may increase insulin sensitivity in liver and muscle via its antioxidant properties. It enhances the uptake of glucose which ultimately results in reduced hyperglycemia. Earlier studies have also revealed that honey has potential to scavenge free radicals. Honey has been reported to ameliorate oxidative stress in pancreas (Fig. 7.15), safeguard the pancreas against oxidative damage, and enhance secretion of insulin which results in better glycemic control (Erejuwa et al. 2012a, b).

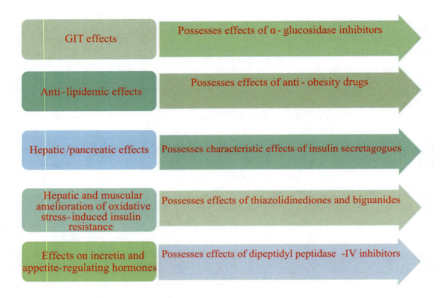

Fig. 7.15 Antidiabetic effect of honey possessing characteristics of antidiabetic drugs

7.2.2 Herbal Honey and Cancer

An unregulated proliferation of cells that may exhibit malignant tendencies is known as cancer. It is the second biggest cause of mortality worldwide, accounting for an estimated 9.6 million deaths in 2018 (Ferlay et al. 2014). Approximately 19.3 million new cases of cancer and nearly 10.0 million deaths due to cancer have been estimated worldwide in the year 2020. Female breast cancer has been found to be the most commonly diagnosed cancer, with an estimated 2.3 million new cases followed by lung, colorectal, prostate, and stomach cancers. Lung cancer was reported to be the main cause of cancer death with an estimated 1.8 million deaths followed by colorectal, liver, stomach, and female breast cancers. The global cancer load is expected to be 28.4 million cases in 2040 which indicates 47% rise from 2020 (Sung et al. 2021). The primary three stages of the development of cancer are initiation, promotion, and proliferation (Fig. 7.16). The initial step of carcinogenesis, called initiation, is marked by an accumulation of altered DNA which involves irreversible genetic damage. The initiation stage is followed by the promotion stage. The promotion stage is characterized by the progression of mutated cells, and alternation in genome of replicated cells results in formation of a benign mass of aberrant cells called as tumor. Next is the proliferation stage, which comprises cancer cells spreading via the lymphatic or circulatory systems to distant locations (tissues and organs).

7.2 Herbal Honey in Non-Communicable Diseases

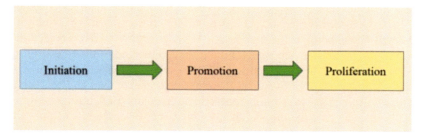

Fig. 7.16 Stages of the cancer development

The most prevalent types of cancer include (Plummer et al. 2016):

Lung cancer (2.09 million cases)
Breast cancer (2.09 million cases)
Colorectal cancer (1.80 million cases)
Prostate cancer (1.28 million cases)
Skin cancer (non-melanoma) (1.04 million cases)
Stomach cancer (1.03 million cases)

7.2.2.1 History of Cancer

Paleopathological evidences suggest that cancer in animals existed in prehistoric times, long before the evolution of humans. In the field of medicine, the Edwin Smith Papyrus, which dates back to around 3000 BC has the first documented account of a disease, mentioning that the breast cancer with a protruding tumor was a serious condition for which there was no effective treatment. The first reference of a fatty tumor, or soft-tissue tumor, is found in the Ebers Papyrus, which dates to around 1500 BC. It also includes cancers of the rectum, stomach, uterus, and skin. The Egyptians used salts, knives, and cautery to cure tumors and cancers. They additionally introduced arsenic paste that was used as "Egyptian ointment" until nineteenth century. During the same period, the Hebrews, Chinese, Indians, Persians, and Sumerians had a preference for herbal cures such as boiling cabbage, figs, tea and fruit juices. However, in more severe cases, they were not afraid to use pastes and solutions made of iron, copper, sulfur, and mercury. For almost 3000 years, several of these mixtures were used both internally and externally in different concentrations. Hippocrates (460–375 BC) and his adherents disagreed that superstitions could be the cause of cancer. They assumed that natural factors are the origin of cancer. They believed that an excess or lack of secretions from the body such as bile, mucus, blood, or other bodily fluids, especially in old age, could cause cancer. Pliny the Roman (AD 23–79) documented cancer remedies in his Materia Medica. Before or after attempting surgery, he recommended using compound herbal remedies and other treatments internally for patients with advanced cancer. His most well-received medicine was a boiling concoction of egg white, honey, powdered falcon poop, and sea crab ash. Aretaeus (AD 81–138) provided the first thorough explanation of the symptoms, signs, and method of treatment for uterine cancer. In his notes, he identified two different types of cancers: one that was firm to the touch and non-ulcerated and the other that smelled bad and was ulcerated. There was groin pain and swelling related to both tumors. Both lesions were considered chronic and fatal by him, but the ulcerated one was more severe and had little chance of healing. He considered uterine hemorrhage in conjunction with uterine hypertrophy to be an irreversible disorder (Fell 1857).

7.2.2.2 Causes of Cancer

Epidemiological data across time shows that environmental factors such as lifestyle choices and particular occupation account for the majority of cancer cases. The involvement of environmental agents in multistep carcinogenesis has been confirmed by experimental evidence, but genetic predisposing factors are still significant drivers. A malignant phenotype develops as a result of complex interactions between internal and environmental variables. In both human and experimental animal models, the complex multistep process can take up more than half of the organism's life. These and other characteristics of the carcinogenic process indicate that the transformation of normal cells into completely malignant tumor cells is mediated by many pathways and multicellular genes. Ionizing radiation, UV light, chemicals, and viruses are examples of known carcinogens. Within a fraction of second, the interaction of radiation with the cell takes place. They must consequently alter the work of cellular genes. DNA modification is the first step in the neoplastic process, and it may be inherited by offspring cells. These changes consist of methylation, rearrangements, amplifications, oncogene activation, and mutations. Agents that may function as promoters alter the irreversible commencement of the transformation process. It has been demonstrated that tumor promoters function at various phases and cause cells to proliferate clonally. It is theoretically possible to intervene at different stages and alter the frequency and course of the neoplastic process because promotion and later progression occur at several stages. Genes within cells express themselves abnormally as a result of initiation and promotion. Permissive elements seem to play a major role in the events that occur. Permissive factors encompass both physiological and genetic susceptibility, including a multitude of growth factors and hormones that control cellular differentiation and play a role in the malignant process. Numerous different oncogenes have been identified as linked to particular cell growth modulators. Growth factors and hormones function on cellular receptors to initiate a series of intracellular biochemical signals that activate and repress different gene subsets. A series of genetic events resulting in the uncontrolled production of growth factors or parts of their signaling pathways cause malignant cells to grow (Weinberg 1989).

Three types of external agents that we consume interact with genetic variables to cause cancer (Blackadar 2016). These substances include the following:

Physical carcinogens: Physical carcinogens include radiation from sources that emit X-rays, gamma, beta and alpha radiation, sunlight's UV rays, uranium, and radon.

Ionizing Radiation

Radiation is inevitable in life. It is found in nature and all around us. It has been utilized for different purposes since it may also be generated artificially. When radiation has enough energy to dislodge electrons from their orbits in the atoms or molecules that make up the irradiated substance, it is referred to as "ionizing" radiation. Chemical bonds are broken as a result, changing things permanently. Ionizing radiations include X-rays, gamma rays, radon, etc. These radiations ionize immediately. They have sufficient energy to damage DNA and cause cancer.

UV Irradiation

The primary cause of skin malignancies, including malignant melanoma, among nonionizing radiations is UV radiation from sunlight which further clarifies why skin cancer rates are higher in southern climates. It has been reported that cumulative (workplace) exposures seem more strongly linked to non-melanoma skin cancer, although intermittent exposure (play and childhood sunburn exposures) might be essential in melanoma.

Chemical carcinogens: They include substances like n-nitrosamines, asbestos, cadmium, benzene, vinyl chloride, nickel, and benzidine. There are also approximately 60 substances that are known to be highly carcinogenic from tobacco use or smoking cigarettes as well as contaminants found in drinking water (arsenic) and aflatoxin in food (Leitch 1923).

Biological carcinogens: They include infections caused by certain bacteria, viruses, or parasites. Examples of pathogens include the human papillomavirus (HPV), the Epstein–Barr virus (EBV), hepatitis B and C, the Markel cell polyomavirus, *Schistosoma* species, and *Helicobacter pylori*.

Pollution: Etiology of lung cancer has long been linked to pollutants in metropolitan atmospheres. Particular concern may be with polycyclic hydrocarbons and ozone, the most common contaminant in metropolitan atmospheres. Human health is seriously threatened by indoor pollution from radon gas and tobacco smoke, as well as outdoor pollution from ozone and hydrocarbons. When certain substances like ozone and hydrocarbons interact with different macromolecules within the cell, free radicals are generated. These mechanisms have been partly linked to the possibility of developing neoplastic disorders.

7.2.2.3 Alcohol Consumption

According to International Agency for Research on Cancer, moderate to heavy alcohol intake is strongly associated with an increased risk of cancer in the oral cavity, pharynx, larynx, esophagus, liver, colon, rectum, and female breast. It has been estimated that in comparison to nondrinkers, the relative risk in case of heavy drinkers is 5.13 for oral and pharyngeal cancer, for esophageal cancer it is 4.95, for laryngeal cancer it is 2.65, for liver cancer it is 2.07, for colorectal cancer it is 1.44, and for breast cancer it is 1.61 (Lopez-Lazaro 2016). Moderate consumption of alcohol (up to one drink per day) raises the chance of developing cancer in the breast, pharynx, esophagus, and oral cavity but not in the liver, larynx, colon, or rectum. An estimated 5.8% of cancer-related fatalities worldwide are thought to be related to alcohol consumption. A major contributing factor to cancer is exposure to mutagenic substances; for instance, smoking increases the risk of pharyngeal and oral cancer by five to ten times. But more than any known mutagenic substance, aging and tissues' ability to regenerate themselves raise the chance of cancer. Exposure to mutagenic chemicals ultimately causes cancer. Ethanol is not mutagenic; however, majority of ethanol consumed is converted by the liver to the mutagenic substance acetaldehyde, which enters the bloodstream and is thought to be the cause of alcohol's carcinogenic properties. But why drinking alcohol raises the risk of some malignancies but not others is still not clear. Alcohol intake is associated with higher levels of estrogens, which may account for the increased risk of breast cancer. Estrogens are crucial in the development of breast cancer (Dorgan et al. 2001).

7.2.2.4 Mechanisms of Carcinogenesis

The somatic mutation theory, which is the currently accepted theory of carcinogenesis, states that DNA mutations and epimutations that cause cancer interfere with the programming governing these orderly processes and upset the normal balance between cell death and proliferation (Majerus 2022). Individuals may be predisposed to cancer by variations in inherited genes. Furthermore, environmental variables including radiations and toxins produce mutations that may aid in the development of cancer. Haphazard errors in regular DNA replication could also lead to mutations that cause cancer. Tumors can result from mutations in multiple crucial genes. The tumor-suppressor gene p53 has mutations in roughly 50% of human tumors. Uncontrolled cell division is made possible by the inactivation of the p53 protein, which protects a cell cycle checkpoint. When a cell divides, mutations resulting from DNA lesions, such as broken chromosomes or damaged bases, are likely to occur. Lesions from an external mutagen increase beyond the background rate of endogenous lesions. The rate at which a given lesion multiplies determines how successful it is at excision by DNA repair enzymes and on the likelihood that it results in a mutation during cell division. This is a crucial component of mutagenesis because a DNA lesion during cell division might result in a point mutation, deletion, or translocation (Cohen et al. 1991; Ames et al. 1993). Therefore, an agent's ability to generate a rise in cell division rates in relevant cells is a key component of its mutagenesis effect. The stem cells, which are not destroyed in contrast to their daughter cells, are the cells that appear to be the most important for cancer. Stem cell division rates that are higher lead to more mutations and ultimately, cancer. Various substances, including elevated levels of specific hormones can lead to accelerated cell division and hence an elevated risk of cancer (Henderson et al. 1982; Moalli et al. 1987).

7.2.2.5 Diagnosis of Cancer

Cancer is diagnosed through the screening tests. Pap tests, mammograms, and colonoscopies are a few examples of the screening test. In order to verify if the body is aberrant, tests like CT, MRI, X-ray, and ultrasound scans are carried out prior to screening tests. The radionuclide test is carried out to visualize area such as inside bones or certain lymph nodes. When a person has cancer but shows no symptoms, their illness is identified during testing for other conditions or problems. If the patient exhibits cancer symptoms, the doctor performs a number of tests.

7.2.2.6 Lab Tests

Urine, blood, and other bodily fluids are tested in lab to determine the quantities of compounds that can cause cancer in the bodies, both at low and high concentrations. In reaction to cancer, both cancer cells and other cells release tumor markers. A doctor must confirm the results of lab tests by conducting additional cancer tests because they do not provide an accurate diagnosis of cancer.

7.2.2.7 Imaging Test

This test creates images of specific internal body parts to assist in determining whether a tumor is present or not. It includes assessments such as:

CT Scan

These scans are used by computer-linked X-ray scanners to produce three-dimensional images of body organs from various perspectives. Usually, prior to scanning, a dye or other contrast agent is ingested which makes it simpler to detect specific body parts in the image. The body is scanned by moving the camera inside the donut-shaped scanner.

MRI

This scan can also be used to build detailed images of the body's organs by capturing photographs of them in slices. To remove the slices, a strong magnet and radio waves are used. The exact distinction between diseased and healthy tissues is visible in this scan. Similar to CT scans, MRIs require the administration of a dye in order to proceed with the imaging. Its a circular chamber contraption that produces loud pounding noises and rhythms when the body is pushed through it.

Pet Scan

The 3D image of the body's organs is created by this scan using radioactive glucose material because cancer cells need much more glucose than healthy cells do.

7.2.2.8 Biopsy

In order to diagnose cancer, a biopsy is a procedure where a medical professional takes a sample of tissue from the patient. A pathologist then performs more testing, examines tissue under a microscope, and records all of the information in a pathology report. Patients are given anesthetic and sedatives prior to biopsy in order to help them relax. Several methods are used to collect the biopsy sample:

(a) **With needle:** Tissue or fluid can be removed from the body with a needle. Spinal taps, prostate biopsies, bone marrow aspirations, and liver and breast biopsies are all done with this technique.
(b) **With endoscopy:** This approach involves inserting a small, illuminated tube called an endoscope via a natural body opening, like the mouth or anus, to inspect inside organs. During an examination, if any aberrant tissue is noticed, then it is removed together with the normal tissue. For example, a bronchoscopy or a colonoscopy.
(c) **Through surgery:** The region containing the aberrant cells is surgically removed. Excisional surgery involves removal of all aberrant cells from the affected area together with a tiny portion of normal cells; in incisional surgery, only a portion of the abnormal area is removed. In case someone is diagnosed with cancer, the physician will determine the optimal treatment based on the cancer's stage.

7.2.2.9 Treatment of Cancer

Depending on the type of cancer and its stage of progression, many forms of treatments are available. While some individuals receive only one cancer treatment, most patients receive many therapies, such as radiation therapy and surgery. Different kinds of treatment are:

7.2 Herbal Honey in Non-Communicable Diseases

Surgery: Surgery may involve the removal of lymph nodes in order to stop or slow down the spreading of disease and eradicate cancer from the body.

Radiation therapy: High radiation dosages are utilized in this therapy to treat cancer through reducing tumor and eliminating cancer cells.

Chemotherapy: Chemicals are employed in this therapy to shrink tumors and destroy cancer cells, but the side effects of chemotherapy are severe. It can be done by the following methods.

(a) **Oral:** as liquids, tablets, and capsules.
(b) **Intravenous:** in the vein directly.
(c) **Intramuscular:** given to the thigh, arm, or hip muscles.
(d) **Intrathecal:** injected between the layers of tissue covering the brain and spinal cord, called an intrathecal space.
(e) **Intra-arterial:** injected directly into an artery.

Immunotherapy: In this form of therapy, medicines or other therapies are used to strengthen the immune system. Immunotherapies are given in different ways:

(a) **Oral:** as capsules or pills.
(b) **Intravenous:** directly into veins.
(c) **Intravesical:** directly into bladder.
(d) **Topical:** when skin cancer is still in its early stages, cream is applied externally to the skin's surface.

Hormone therapy: This type of treatment uses hormones to block and slow down the growth of cancers including breast and prostate cancer.

Stem cell transplants: Using this treatment, cancer patients' stem cells which are killed by extremely high radiation or chemotherapy doses are restored.

Effects of cancer treatment: Normal cells, tissues, and organs may also get affected by cancer treatment. Side effects of the treatment are manifestations of the therapeutic effect of the medication. The common negative impacts are listed in Table 7.1.

Table 7.1 Various side effects of cancer treatment

S. no.	Part of body treated	Side effects of the treatment
1.	Brain	Tiredness, loss of hair, vomiting and nausea, changes in skin, headache, and blurry vision
2.	Chest	Weakness, hair loss, skin alterations, changes in the throat such as swallowing, coughing, and shortness of breath
3.	Head and neck	Fatigue, hair loss, changes in the mouth, skin, taste, throat including trouble in swallowing, reduced thyroid gland activity
4.	Pelvis	Weakness, tiredness, loss of hair, vomiting and nausea, male sexual problems, male fertility issues, issues with sex (women), fertility issues in women, skin alterations, changes in the bladder
5.	Stomach and abdomen	Fatigue, hair loss, diarrhea, vomiting, nausea, skin alterations, changes in the bladder
6.	Rectum	Fatigue, hair loss, diarrhea, men's fertility issues, women's sexual issues, skin problems

7.2.2.10 Effect of Honey and Herbs on Cancer

Natural honey has been used for food and medicine since ancient times. After in-depth investigation of the literature, it has been discovered that raw honey is the most effective sweetener utilized in earlier times and it has been widely used several million years ago. Due to its therapeutic properties containing several ingredients, natural honey has been ingested in large quantities. In addition to fructose and glucose, it also contains numerous amino acids, vitamins, minerals, and enzymes. Natural honey possesses anticancer, antioxidant, antimutagenic, antibacterial, and wound healing activities. Its ability to serve as a defense mechanism has been confirmed because of its phenolic content (Erejuwa et al. 2014). Different types of tissues and cancer cell lines, including the colon, breasts, endometrial, prostate, kidney, oral and cervical cancers have been used to evaluate anticancer properties of honey. Chemotherapies like 5-fluorouracil and cyclophosphamide are stimulated by raw honey. The antioxidant and anticancer properties of honey are attributed to its polyphenol content which has been demonstrated through a variety of analytical methods including DPPH (Diphenyl-1-picrylhydrazyl), FRAP (Ferric Reducing Antioxidant Power), ORAC (The Oxygen Radical Absorbance Capacity), ABTS [2, 2-azinobis (3ehtylbenzothiazoline-6-sulfonic acid) diammonium salt], and TEAC [6-hydroxy-2, 5, 7, 8-tetramethylchroman-2-carboxylic acid (Trolox)-equivalent antioxidant capacity] (Moniruzzaman et al. 2012).

7.2.2.11 Anticancer Mechanisms of Honey

Some potential processes by which natural honey may prevent cancer include cell cycle arrest, permeabilization of outer mitochondrial membrane, the induction of apoptosis, the modulation of insulin signaling and oxidative stress, the amplification of inflammation, and introduction of estrogenic activity (Fig. 7.17, Table 7.2).

7.2.2.12 Cell Cycle Arrest

The cell cycle which consists of the four successive segments G1, S, G2, and M phase is a series of synchronized events where cell development and propagation are strongly recorded. At the M phase, cells divide into duplicate progeny cells and at the S phase, DNA replication occurs. In response to external signals, the cells at the G1 stage undergo consecutive mitosis or are removed from the cell sequence into a dormant phase known as G0. Events throughout the cell cycle are regulated by a cascade of protein kinases and checkpoints. Dysregulation of the cell cycle results in the growth of frenzied cancer cells (Pichichero et al. 2010). One potential strategy to reduce tumor growth is through cell cycle arrest caused by honey therapy. According to certain theories, honey and its components such as flavonoids and phenolics are essential to stop colon cancer, glioma, and melanoma cell lines in the G1/G0 stage of their cell cycles. Numerous investigations, such as 3-(4,

7.2 Herbal Honey in Non-Communicable Diseases

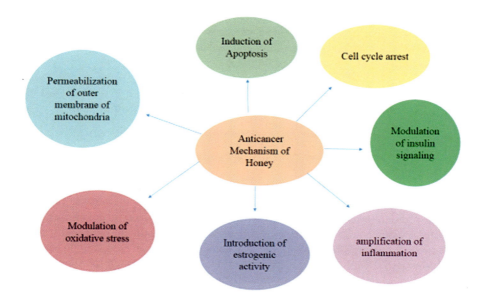

Fig. 7.17 Anticancer mechanism of honey

Table 7.2 Effects of honey on the suppression of tumor and cancer cells

Sr. no.	Types of cancer	Effect of honey
1.	Human liver cancer	Controls angiogenesis, triggers apoptosis, prevents DNA damage from mutagenic agents, and inhibits the growth of cells
2.	Human breast cancer	Suppresses the function of estrogen, stops cell division, triggers apoptosis, and lowers the potential of the mitochondrial membrane
3.	Human colorectal cancer	Reduces inflammation, decreases mitochondrial membrane potential, increases production of reactive oxygen species (ROS), depletes intracellular non-protein thiols, causes DNA damage, and inhibits cell proliferation, apoptosis, and cell cycle arrest
4.	Human prostate cancer	Triggers apoptosis and inhibit cell proliferation
5.	Human bladder cancer	Inhibition of cell proliferation
6.	Human kidney cancer	Triggers apoptosis
7.	Human oral cancer	Inhibition of cell proliferation
8.	Human bone cancer	Inhibition of cell proliferation
9.	Human skin cancer	Inhibition of cell proliferation and cell cycle arrest

5-dimethylthiazol-2-yl)-2, 5-diphenyl tetrazolium bromide (MTT) assay and the trypan blue exclusion experiment have demonstrated that the antiproliferative activity of honey is dose- and time-dependent (Yang et al. 2013). Recent findings confirm that honey's lethal effect on non-small cell lung cancer cells (NCIeH460) is mediated by cell cycle arrest at the G0/G1 phase. Another experiment has shown that chrysin, a significant component of honey, inhibits the growth of human and murine melanoma cells by arresting the cell cycle at the G0/G1 phase (Afroz et al. 2016).

7.2.2.13 Induction of Mitochondrial Pathway and Permeabilization of the Outer Membrane

Radiotherapy and chemotherapy both cause cell death through the essential mechanism of mitochondrial pathway activation. The inner membrane space of mitochondria contains cytochrome-C, which is one of the proteins released by the intrinsic mitochondrial pathway. Other stimuli that cause the cell to die include oxidative stress, physical stress, and nutrition (Ren et al. 2012). Because it releases cytochrome-C upon activation of the mitochondrial intrinsic pathway, honey with a high flavonoid content has been suggested as a potential cytotoxic mediator (Jaganathan and Mandal 2010). The process of triggering cell death through the induction of mitochondrial outer membrane permeabilization, or MOMP, results in the leakage of mitochondrial proteins into the cytosol, which is a typical mechanism involving mediators with anticancer effects. It has been reported that Indian honey causes MOMP in HT-29 and HCT-15 colon cancer cells by lowering mitochondrial membrane potential. Research revealed that MOMP is caused in many cell lines by quercetin, a subclass of flavonoids found in honey. Therefore, cell death attributable to honey and its flavonoid components involves MOMP (Nassar and Blanpain 2016).

7.2.2.14 Induction of Apoptosis

Cancer cells exhibit two characteristics: uncontrollably proliferating cells and uneven apoptotic turnover. Apoptosis inducers are drugs that are typically used to treat cancer. Three phases can be distinguished in apoptosis, also known as programmed cell death: (a) commencement phase, (b) proliferation phase, and (c) progression or deprivation stage. Through death-inducing signals, the initiation step promotes the transduction cascade of proapoptotic signals. One method of bringing cell death with the commitment to the effector stage is through the mitochondria, a crucial regulator. The final deprivation phase includes cytoplasmic and nuclear processes. Nuclear alterations include chromatin and nuclear contraction, cell shrinkage, genetic material disintegration, and tissue blebbing. Several types of tumor cells undergo apoptosis when honey depolarizes the mitochondrial membrane. While natural honey has been shown to downregulate the antiapoptotic protein Bcl-2, it has been shown to increase caspase 3 and other proapoptotic proteins. Application of honey containing aloe vera has been shown to decrease Bcl-2 expression and enhance the appearance of the proapoptotic protein Bax in Wistar rats. Caspases 9 is stimulated first, which then activates the executor protein caspase 3 (Yaacob et al. 2013).

7.2.2.15 Modulation of Oxidative Stress

Reactive oxygen species (ROS) and oxidative stress have a pivotal role in the suppression and development of cancer cells (Balestrieri et al. 2012). The idea that ROS have two functions in cancer, i.e., an inhibitory role and a stimulatory role is supported. ROS in trace quantities promote cell proliferation. Furthermore, increased ROS intensities, which are the foundation of oxidative damage, have been linked to a number of cancer types, including breast, stomach, colon, and lung cancers which are associated with the effect of honey in regulation of oxidative stress (Peddireddy et al. 2012).

7.2.2.16 Anti-inflammatory and Immunomodulatory Effect

Honey possesses anti-inflammatory response as it reduces inflammation when applied in cell cultures. The involvement of honey and its contents in the regulation of proteins such as COX-2, iNOS, tyrosine kinase, and ornithine decarboxylase has been recognized (Attia et al. 2008). Non-sugar components of honey may be responsible for its immunomodulatory and immunopotentiating properties (Alam et al. 2014). Furthermore, nigero-oligosaccharides (NOS), a sugar found in honey have been shown to have immune-stimulating properties.

7.2.2.17 Modulation of Insulin Signaling

According to epidemiological research, the main risk factors for the development of various malignancies include obesity, insulin resistance, and type 2 diabetes mellitus. The role of insulin receptors (IR) in carcinogenesis has been confirmed with the evidences that how IR targeted by small compounds is efficient against the reduction of lung cancer cell proliferation (Tandara and Mustoe 2004). P13K/Akt is another crucial element of insulin signaling, as it modifies a number of variables that control both the growth and development of the cell cycle. Impact of honey on the Akt-activated insulin signaling pathway under hyperglycemia settings was recently demonstrated in HIT-T15 cells that the development of insulin resistance entails elevated levels of Nf-kB, insulin receptor substrate 1, and MAPK. Quercetin and gelam honey extract pretreatment has been shown to improve insulin content and insulin resistance. Honey eating may be used to modify insulin signaling in cancer cases (Erejuwa et al. 2012a, b).

The use of contemporary facilities to comprehend their biological activity and chemical components has greatly aided in the streamlining of several challenges related to serious illnesses. The existence of phytochemicals, which are categorized as main and secondary substances, allows plants to exhibit such medicinally significant qualities. However, secondary compounds provide a wide range of health benefits including anticancer properties. Examples of these compounds are coumarins, alkaloids, terpenoids, and phenolic compounds (Krishnaiah et al. 2007). The

Catharanthus roseus plant alkaloids are well-known for their sedative, hypotensive, and anticancer properties. Vincristine and vinblastine are two of the most potent and readily accessible compounds identified from *C. roseus*. Vinblastine and vincristine are recognized for their ability to block mitosis, indicating that they prevent cell division and ultimately lead to cell death. The formulation of vincristine aids in adhering to the tubulin protein and prevents the cells from splitting their chromosomes during metaphase, which is responsible for cell death (Jordan 2002). The growths of several human cancers are controlled by these alkaloids of *C. roseus*. Vinblastine is advised for the treatment of Hodgkins disease and choriocarcinoma. Vincristine, the other alkaloid, is used to treat childhood leukemia. While vincristine is marketed as oncovin, vinblastine is sold under the brand Velban on the market. Cancer can also be prevented and treated using turmeric. The ability of curcumin to suppress growth in a range of tumor cell types has been linked to its anticancer potential. More than 30 different protein targets are bound by curcumin: transcript factors (NF-kB and activator protein-1), growth factor receptors (EGFR, human epidermal growth factor receptor 2 (HER2)], kinases (MAPK, PKC, and protein kinase A (PKA)), inflammatory cytokines (TNF, interleukins), cell cycle-related proteins, matrix metalloproteinases (MMPs), and urokinase plasminogen activators (Goel et al. 2008). *Panax ginseng*, or ginseng root, has long been used as a medicine to treat a variety of illnesses, including cancer. Ginseng has a variety of ingredients including vitamins, volatile oils, ginsenosides, polysaccharides, and amino acids. Of these, ginsenosides and polysaccharides have an anticancer impact (Nag et al. 2012).

7.2.3 Herbal Honey and Wound Management

Any disturbance or injury to living tissue, including skin, mucous membranes, or organs is referred to as a wound. Injuries can arise abruptly from mechanical, thermal, or chemical assault. They can develop gradually over time as a result of underlying medical conditions such diabetes mellitus, venous or arterial insufficiency, or immunologic disorders (Kujath and Michelsen 2008). The location, mechanism, depth, start timing (acute vs. chronic), and sterility of the wound, among other variables, can all significantly affect how a wound appears. Our skin is a highly adapted, multifunctional organ that has evolved over millennia of evolution in order to protect us from the everyday assault of chemicals, physical stresses, and UV radiation. Our skin frequently sustains damage from the harsh environment. It has advanced healing mechanisms that enable rapid and effective skin healing. Even while people have a significant natural capacity for healing, there are several cellular components of the damage response that can weaken, which can affect how well wounds heal. This attenuation is most typically a result of pathogenic systemic alterations, such as those associated with elderly age or uncontrolled diabetes. It is clear that the two main risk factors for developing a chronic wound are age and diabetes. Unfortunately, there is a considerable medical requirement unfulfilled in the domain of these chronic wounds, which are mostly pressure sores, diabetic foot ulcers, and venous ulcers. These wounds are also becoming more prevalent worldwide. The use of cutting-edge research technologies will be crucial to understanding the underlying cellular and molecular mechanisms of both pathological and acute repair (Wilkinson and Hardman 2020).

7.2.3.1 Mechanism of Wound Healing

In addition to serving as an interface with the environment around us, our skin performs a number of vital homeostatic tasks, such as controlling thermostability and detecting external stimuli. The skin serves as the body's first line of defense, shielding internal tissues from photic, chemical, mechanical, and thermal damage as well as desiccation (Takeo et al. 2015). This defense includes a highly developed immune response that fends off harmful infections and sustains commensal bacteria through a host–microbiota axis that has been exquisitely regulated (Naik et al. 2015). Additionally, the skin has developed quick and effective mechanisms to seal up damage to its barrier, a process known as the wound healing response. The four basic stages of wound healing are hemostasis, inflammation, proliferation, and dermal remodeling (Broughton et al. 2006).

7.2.3.2 Stages of Wound Healing

Hemostasis

Damaged blood vessels quickly constrict after injury, forming a blood clot that stops bleeding due to vascular damage. When platelets come into contact with the vascular subendothelial matrix, they become activated. Platelets are the primary factor in hemostasis and coagulation. Platelet receptors, such as glycoprotein VI, interact with collagen, fibronectin, and von Willebrand factor, among other extracellular matrix (ECM) proteins, to enhance the blood vessel wall's adhesion to the platelets. Following this, thrombin stimulates platelet activation, causing a conformational shift and the release of dense and alpha granules carrying bioactive chemicals that strengthen coagulation. In order to plug the wound and stop the bleeding, an insoluble clot comprises fibrin, fibronectin, vitronectin, and thrombospondin forms (Zaidi and Green 2019). In addition, the clot performs a variety of secondary roles, including as protecting against bacterial invasion, acting as a scaffold for incoming immune cells and housing a reservoir of growth factors and cytokines that influence the behavior of injured cells during the early stages of healing (Delavary et al. 2011). Through their ability to either directly capture immune cells in the clot or upon degranulation, platelets play a critical role in the recruitment of immune cells to the injury site (Golebiewska and Poole 2015). Platelets are the most prevalent cell type during the early stages of healing, and they actively contribute to the early suppression of bacterial infection. Several toll-like receptors (TLRs) are expressed by them, and these TLRs control the synthesis of antimicrobial peptides. The coagulation process is stopped after a large enough clot has formed, preventing excessive thrombosis. Prostacyclin inhibits platelet aggregation, antithrombin III inhibits thrombin, and activated protein C degrades coagulation factors V and VII. In parallel, smooth muscle and endothelial cells that multiply in response to released platelet-derived growth factor (PDGF) heal the damaged vessel wall (Kingsley et al. 2002).

Inflammation

The primary line of defense against the invasion of pathogenic wounds has developed into innate inflammation. Damage-associated molecular patterns (DAMPs) which are released by necrotic cells and damaged tissue, as well as pathogen-associated molecular patterns (PAMPs), which are produced by bacteria, are what trigger this immune response (Wilkinson and Hardman 2020). By attaching to pattern recognition receptors on resident immune cells, such as mast cells, Langerhans cells, T cells, and macrophages, these PAMPs and DAMPs trigger subsequent inflammatory pathways (Chen and DiPietro 2017). Circulating leucocytes are drawn to the site of injury by the subsequent production of pro-inflammatory cytokines and chemokines. Pro-inflammatory chemicals also cause vasodilatation, which promotes neutrophil and monocyte adherence and diapedesis in conjunction with the production of endothelial cell adhesion molecules such as selectins. Arriving early after injury, neutrophils are drawn into the wound from damaged arteries by chemoattractants such as lipopolysaccharide (LPS), TNF-α, and interleukin 1 (Kolaczkowska and Kubes 2013). Bacterial endotoxins also attract neutrophils and neutrophils and release their own cytokine in response to pro-inflammatory signals and activation of inflammatory signaling pathways (e.g., NF-κB). By phagocytosing necrotic tissue and releasing reactive oxygen species (ROS), antimicrobial peptides, eicosanoids, and proteolytic enzymes, neutrophils eliminate infections and necrotic tissue. Numerous internal and external factors can influence the complicated inflammatory response. Excessive and uncontrollably high inflammation accelerates tissue damage and slows recovery. Pro-inflammatory stimuli like LPS and interferon-gamma (IFN-γ) stimulate classically activated macrophages, which in turn promote inflammation by producing reactive oxygen species (ROS), inflammatory cytokines like IL-1, IL-6, and TNF-α, and growth factors like VEGF and PDGF. Numerous growth factors are also released by anti-inflammatory macrophages to encourage angiogenesis, fibroplasia, and re-epithelialization (Delavary et al. 2011).

Proliferation

In order to coordinate wound closure, matrix deposition, and angiogenesis, keratinocytes, fibroblasts, macrophages, and endothelial cells are extensively activated during the proliferative phase of healing. Changes in mechanical tension and electrical gradients as well as exposure to hydrogen peroxide, infections, growth factors, and cytokines can activate keratinocytes as early as 12 h after injury. Keratinocytes at the wound edge undergo a partial epithelial–mesenchymal transition as a result of this activation, during which they take on a more invasive and migrating phenotype. Re-epithelialization is the process by which leading-edge keratinocytes migrate laterally across the wound to reconstruct the epidermal layer (Shaw and Martin 2016).

Remodeling

It is the final stage of wound healing where the maturation of granulation tissue into scar involves reduction in the number of capillaries through aggregation into larger vessels and a decrease in the amount of glycosaminoglycans (GAGs). Organization of collagen also alters, enhancing the tensile strength of the tissue. Tensile strength is also increased due to cross-linking of collagen by lysyl oxidase enzyme secreted by fibroblasts in extracellular matrix (Schultz et al. 2011).

7.2.3.3 Current Therapies for Wound Management

The wound healing process involves assessing the cause of the wound and treating systemic and lifestyle factors from the perspective of the patient. Debridement or the surgical removal of necrotic, diseased, or hyperkeratotic tissue is frequently the first step in local therapy for diabetic foot ulcers. Removing the chronic tissue to the less damaged epidermis is believed to initiate the usual reparative healing pathways while inducing an acute injury reaction (Stephen-Haynes and Thompson 2007). After an application of a customized dressing, wounds are irrigated with saline or an antibiotic solution. Modern dressings are made of a variety of materials with recognized pro-healing or antibacterial qualities to help in tissue restoration. More sophisticated options exist like constantly developing negative pressure wound therapy technique. Even though there are many therapies available, the best practice of wound management relies primarily on patient's compliance and virtually solely addresses secondary causes of chronicity. A key factor to chronic wound recalcitrance is persistent, antibiotic-resistant biofilm infection. Therefore, the discovery of innovative antibiotic and anti-biofilm medicines has occupied a significant amount of recent wound research. While newer formulations of non-antibiotic antimicrobials like silver salts are cytotoxic to the host, they reduce the load of bacteria. Nanoparticles may accelerate the healing of wounds and have less cytotoxicity. New antimicrobial therapies like bioactive glass and cold atmospheric plasma may also be helpful in tissue healing (Xie et al. 2019). The majority of antimicrobials have wide actions and don't specifically target any one pathogenic species or strain. This is significant because commensal microorganisms benefit the skin upkeep and healing of wounds (Lukic et al. 2017). Additionally, commensal biofilms in diabetic wounds do not result in persistently delayed healing, in contrast to their pathogenic counterparts. Because of this, more focused antimicrobial therapies, such as phage therapy or pharmaceutical suppression of bacterial pathogenicity processes like quorum sensing may be more effective and particular (Johnson and Abramovitch 2017). Moreover, the majority of therapies concentrate on the bacterial aspect of infection; however, the variety of fungi present in wounds is also associated with the degree of healing. In order to clarify the function of host–microbe interactions in pathological repair, future studies should consider the whole ecosystem around the wound (Kalan et al. 2016).

7.2.3.4 Effect of Honey and Herbs on Wound Management

Honey is a magnificent gift from nature regarded as a superior food and a wellspring of conventional medicine produced by honeybees. It has been shown in numerous trials to be beneficial for treatment of different ailments including respiratory infections, diabetes, skin infections, wounds, and digestive disorders. Honey's involvement in wound healing is best described as preventing and limiting bacterial infection which lowers the wound's bioburden. This function was initially thought to be derived from biochemical characteristics related to the production of peroxide through intrinsic glucose oxidase activity. According to more recent studies, infection can be controlled even when catalase is present which supports the concept that there are other non-peroxide-mediated mechanisms. It has been found that *Staphylococcus aureus* colonies treated with manuka honey exhibited cell division arrest, indicating that honey may be able to hinder the progress of the bacterial cell cycle (Henriques et al. 2010). The hygroscopic properties of honey are thought to restrict the amount of wound edema and inhibit the growth of biofilms, which in turn contributes to antibacterial action (Alandejani et al. 2009, Okhiria et al. 2009). Effect of honey to control the inflammatory phase is most strongly supported by various studies.

Anti-inflammatory activity of honey was first demonstrated in guinea pigs (Church 1954). Furthermore, studies have demonstrated that honey shortens the overall wound healing period while enhancing tissue granulation and epithelialization during the proliferative phase (Galiano and Mustoe 2007). Other roles of honey in the physiology of the wound have also been proposed. Manuka honey dressings, for example, have been demonstrated to reduce the alkaline pH of chronic wound bed which is linked to better healing results (Schneider et al. 2007).

Aloe vera, commonly known as Ghrit Kumari, is known to contain over 100 active ingredients, many of which are astringent, hemostatic, antidiabetic, antiulcer, antiseptic, antibacterial, anti-inflammatory, antioxidant, anticancer, and antidiarrheal. *Aloe vera* also has the ability to heal wounds. On an excision wound model, *Aloe vera* leaf pulp has a significantly superior and quicker capacity to cure wounds than povidone iodine ointment (Purohit et al. 2012). *Aloe vera* has also reportedly been shown to prevent microbial infections on the damaged surface in addition to accelerating the healing process. According to reports, *Aloe vera* increases the amount of lysyl oxidase, which cross-links newly created collagen and the rate at which collagen turnover occurs (Chithra et al. 1998). Applying *Aloe vera* gel topically greatly accelerates wound healing and contraction. Collagen serves as the precursor protein for wound healing, and its level is greatly impacted by it. In the latter stages of the wound healing process, *Aloe vera* gel has been shown by histological investigations to expedite epithelialization, neo-vascularization, and enhanced wound contraction. Its ability to cure has been attributed to glucomannan, a polysaccharide-containing substance. This substance increases the production and release of collagen via influencing fibroblast growth factor, which in turn impacts the activity and proliferation of these cells. Mucilage of *Aloe vera* also promotes transversal connections between collagen bands without altering the structure of the collagen, accelerating the healing process.

By actively activating many immunological systems including antibody generation, the release of mediators of hypersensitive reactions, and the tissues' reaction to these mediators at the target areas, *Ocimum sanctum* plays a crucial part in the healing process of wounds (Prakash and Gupta 2005). The volatile oil found in *Ocimum sanctum* leaves is made up of limonene, borneol, copaene, caryophyllene, and elemol. Other compounds that aid in wound healing include phenolic compounds (rosmarinic acid, apigenin, cirsimaritin, and isothymusin), flavonoids (orientin, vicenin), and aromatic compounds such as methyl chavicol and methyl eugenol (Pattanayak et al. 2010). Dinkum oil is the oil that is extracted from *Eucalyptus globules* by steam distilling fresh leaves. It is applied to wounds in skin care. Both topical and oral application of *Eucalyptus citriodora* extracts prove to be highly efficacious in the treatment of cutaneous wounds. These extracts quicken every stage of wound healing. The mechanism of action of these concentrates has been suggested to involve wound contraction during the proliferative stage, angiogenesis, collagen deposition, granulation tissue development, and epithelization. These activities are attributed to the extract's phytoconstituents which include tannins, flavonoids, and phenolic compounds working synergistically. Active chemicals such as nimbidin, nimbin, and nimbidol contained in *Azadirachta indica* (neem) have anti-inflammatory and antimicrobial activity which help in expediting the wound healing process. Neem also has a high content of vitamins, minerals, and amino acids, all of which are important for the proliferation stage of the healing process of wounds (Subapriya and Nagini 2005). The chemical component of turmeric (*Curcuma longa*) called curcumin has been shown to have important wound healing properties in addition to anti-infective, antioxidant, anti-inflammatory, antimutagenic, anticarcinogenic, and anticoagulant characteristics. It accelerates the healing process by acting on several phases of the wound healing process. Additionally, curcumin can improve collagen deposition, granulation tissue production, tissue remodeling, and wound contraction (Akbik et al. 2014). Curcumin stimulates the production of growth factors that aid in the healing process, which expedites the treatment of wound contraction. *Aegle marmelos* leaf extracts have strong antioxidant properties and function as an antigenotoxicant to promote wound healing (Arunachalam et al. 2012). The active ingredients in *Aegle marmelos* root extract speed up wound healing and offer the healed area breaking strength. *Aegle marmelos* fruit pulp has the ability to cure wounds by reducing inflammation and increasing collagen determinants (Gautam et al. 2014).

7.2.4 Herbal Honey and Alzheimer's Disease

Alzheimer's disease is a slowly progressing illness of the brain that mostly impacts memory, cognition, and behavior. Dementia is defined as a decline in mental ability that is severe enough to interfere with routine life, and it is the most prevalent cause of dementia. Alzheimer's disease is a complex and devastating condition, and understanding its various aspects is crucial for both patients and their caregivers. It predominantly affects older adults, and its exact cause is not yet fully understood. Plaques of beta-amyloid and tau tangles are two abnormal deposits of proteins that cause the diseases to build up in the brain and cause disruptions to normal brain function. Dr. Alois Alzheimer, a German psychiatrist and neurologist who first recognized the condition in 1906, is remembered by the disease's name. He described the circumstances surrounding a woman by the name of Auguste Deter, who experienced memory loss, linguistic problems, and behavioral disturbances. After she passed away, Dr. Alzheimer studied her brain and found abnormal protein buildups that are now known as amyloid plaques and neurofibrillary tangles and are indicative of Alzheimer's disease. It's important for one to understand that each person will experience these symptoms differently, including in terms of severity and appearance orders. Here are detailed descriptions of common Alzheimer's symptoms. Initially, people may have trouble recalling recent conversations or occurrences, lose things, or forget names. In advanced Alzheimer's, long-term memory is also affected, and individuals may not recognize close family members or recall their own personal history. Vocabulary and word-finding difficulties may emerge, leading to pauses in speech and difficulty finding the right words. Communication becomes increasingly challenging. Individuals may have trouble following or joining conversations, and they might repeat themselves frequently. Eventually, individuals may lose the ability to speak altogether or may only be able to utter a few words or phrases. People with Alzheimer's often become disoriented in time and place. They may not know the current date or the location of their own home. A number of methods, including herbal treatments like herbal honey, have been investigated to perhaps ease some of its signs and symptoms and enhance the general standard of life for affected people. Herbal honey refers to honey infused with various medicinal herbs, which may have potential benefits for individuals with Alzheimer's disease. Honey has been used throughout history for various medicinal purposes, and there is some evidence to suggest that it may have potential benefits for brain health and conditions like Alzheimer's disease. The use of honey in traditional medical practices, such as Ayurveda and traditional Chinese medicine, has a long history. It was often used to treat various ailments, including those related to cognitive function and memory.

7.2.4.1 Cause of Alzheimer's

Genetics: Alzheimer's disease is greatly impacted by genetic factors. The two primary types of Alzheimer's disease are sporadic Alzheimer's disease and familial Alzheimer's disease (FAD). FAD is often passed on through families that have a history of the condition, and it is brought on by alterations in specific genes, including the APP, PSEN1, and PSEN2 genes. Numerous genetic risk factors, such as the APOE gene, affect sporadic Alzheimer's disease, which is more prevalent. Alzheimer's disease is more likely to affect people with specific APOE gene variants, especially APOE 4.

Vascular factors: According to certain research, vascular health may have an impact on Alzheimer's disease. Diseases like hypertension, diabetes, and high cholesterol levels can increase a person's chance of getting Alzheimer's through their detrimental impact on blood circulation to the brain.

Environmental variants: Despite the importance of inheritance, environmental variables can potentially increase the risk of Alzheimer's disease. These variables could include exposure to certain chemicals, head trauma, and lifestyle decisions including nutrition, exercise, and education.

Age: Age is the primary risk factor for Alzheimer's disease. The risk increases dramatically after age 65, beyond which the illness's rate about doubles every 5 years.

Hormonal changes: Studies investigated at hormone changes as major risk factors for Alzheimer's disease, particularly in postmenopausal women. Researchers have looked into how estrogen protects against damage to the brain.

Lifestyle factors: A healthy lifestyle can reduce the risk of Alzheimer's disease. Numerous factors, such as regular exercise, a balanced diet, mental stimulation, social interaction, and the management of chronic medical illnesses, can have an impact on brain health.

7.2.4.2 Signs and Symptoms

Amyloid plaques: One of the main signs of Alzheimer's disease is the accumulation of beta-amyloid plaques in the brain over time. These unusual protein accumulations can obstruct brain cell interactions and result in cell death.

Tau protein tangles: The development of abnormal tau protein-containing neurofibrillary tangles is another important pathogenic aspect of Alzheimer's disease. These tangled structures prevent neurons from growing and functioning normally.

Neuroinflammation: The development of Alzheimer's disease is hypothesized to be influenced by chronic brain inflammation. Microglias, which are immune cells, can overexert themselves and harm neurons in the brain.

Oxidative Stress: When the body's capacity to combat reactive oxygen species and the production of free radicals are out of balance, oxidative stress results. Oxidative stress can injure brain tissue and have a role in Alzheimer's disease progression (Fig. 7.18).

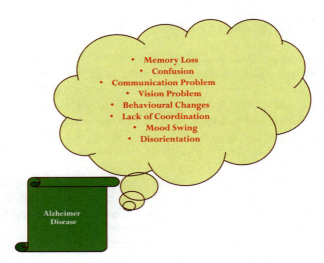

Fig. 7.18 Signs and symptoms of Alzheimer's disease

7.2.4.3 Therapeutic Potential of Honey in Managing Alzheimer's Disease

Antioxidant properties: Some studies have suggested that honey, including herbal honey, possesses antioxidant properties. When cells are exposed to oxidative stress, which has been linked to the onset and progression of Alzheimer's disease, antioxidants are molecules that can help protect the cells. These antioxidants might protect the nervous system.

Anti-inflammatory effects: Alzheimer's disease tends to be related to chronic brain inflammation. Some herbal honey could have anti-inflammatory properties, which contribute to reducing inflammation of neurons and its negative effects on brain health.

Neuroprotective potential: Studies have been conducted to investigate the possible neuroprotective benefits of several honey constituents, including phenolic chemicals and flavonoids. These substances support the survival of neurons and shield them from harm.

Improved cognitive function: Some limited studies have explored the effects of honey, including herbal honey, on cognitive function. While these studies have shown some promise, they are often small-scale and may not provide conclusive evidence of honey's effectiveness in treating Alzheimer's disease or improving cognitive function (Fig. 7.19).

Here are some herbs often associated with herbal honey and their potential effects on Alzheimer's disease:

Turmeric: An active component of turmeric called curcumin has both anti-inflammatory and antioxidant effects. According to some studies, curcumin may help prevent the brain's accumulation of beta-amyloid plaques, which is a symptom

7.2 Herbal Honey in Non-Communicable Diseases

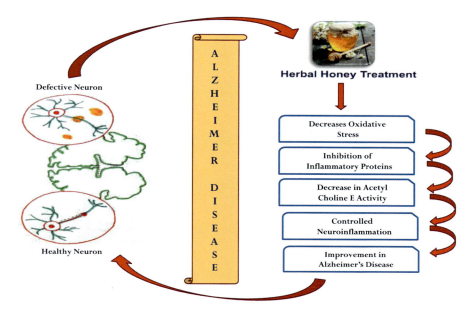

Fig. 7.19 Management of Alzheimer's disease through herbal honey for Alzheimer's disease

of Alzheimer's disease. When combined with honey, turmeric can deliver a relaxing and possibly beneficial combo.

Ginkgo biloba: Ginkgo biloba is an herb known for its potential cognitive-enhancing properties. Some research indicates that it may improve blood flow to the brain and protect against neuronal damage. Combining ginkgo biloba extract with honey could provide a pleasant way to consume this herb.

Rosemary: Rosemary is an herb that contains compounds like rosmarinic acid, which have antioxidant and anti-inflammatory properties. Some studies have suggested that rosemary may help improve memory and cognitive function.

Sage: Sage is another herb with potential cognitive benefits. It contains compounds like rosmarinic acid and acetylcholine, which may improve memory and cognitive function. Sage-infused honey could offer a flavorful way to incorporate this herb into your diet.

Conclusion: Alzheimer's disease is a multifaceted condition with complex contributing factors, including genetics, environmental influences, and age. Herbal honey, infused with various medicinal herbs, shows promise in its potential to offer antioxidant, anti-inflammatory, and neuroprotective benefits for managing some aspects of the disease. While research in this area is ongoing, herbal honey presents a holistic approach to supporting brain health. However, it should be considered as a complementary therapy alongside evidence-based treatments, and further studies are needed to determine its full effectiveness in enhancing the quality of life for individuals with Alzheimer's disease.

7.2.5 Herbal Honey and Gastrointestinal Disorders

Throughout history, illnesses that have caused various symptoms such as discomfort, nausea, vomiting, bloating, diarrhea, constipation, or trouble passing food or feces have been associated with structural digestive tract ailments. The term functional gastrointestinal disorders (FGIDs) refers to a broad category of long-term illnesses for which there is no recognized structural or biochemical cause for the symptoms. The conditions mentioned earlier hold significance for public health due to their very high prevalence, potential for disability, and significant financial impact. Many other illnesses may also occur, but the most well-known of them is irritable bowel syndrome (IBS), which was identified by the Rome III Committee based on a distinctive symptom profile (Nyrop et al. 2007). In Western populations, functional dyspepsia and irritable bowel syndrome (IBS) are two examples of functional gastrointestinal disorders (FGID) that are quite common. IBS is characterized by recurrent abdominal pain or discomfort that lasts for at least 3 days per month and is accompanied by two or more additional symptoms such as improved bowel movements. The onset of IBS symptoms is linked to changes in both the frequency and form of stools. IBS is classified into four subgroups based on the predominant stool pattern as determined by the Bristol stool-form scale: IBS that is primarily constipated, IBS that is primarily diarrheal, mixed, and unclassified (Thompson et al. 1999). There is increasing evidence that inflammation plays a role in the development of both IBS and functional dyspepsia.

7.2.5.1 Causes of the Gastrointestinal Disorder

FGIDs may be explained as a result of dysregulation of brain-gut neuroenteric systems. In addition to having a number of effects on gastrointestinal, endocrine, immunological, and behavioral functions in humans, the brain-gut neurotransmitters linked to these symptoms are not site-specific. The idea that the central nervous system (CNS) plays a role in motility regulation is supported by the facts that (1) motility disturbances associated with irritable bowel syndrome (IBS) depart away when a person sleeps; (2) the frequency of the migrating motor complex (MMC) decreases and propagating velocities increase progressively with alertness and arousal. (3) Patients with IBS have a different electroencephalography sleep pattern than healthy subjects; and (4) patients with IBS have a different CNS response to rectal distension than controls. Increased motor reactivity, enhanced visceral hypersensitivity, altered mucosal immune and inflammatory function including alterations in bacterial flora and altered CNS-enteric nervous system (ENS) regulation are some of the physiological determinants linked to the symptoms of FGIDs. More complicated, IBS (category C1) is caused by a confluence of factors including visceral hypersensitivity, mucosal immunological dysregulation, changes in bacterial flora, dysmotility, and CNS-ENS imbalance (Drossman and Dumitrascu 2006).

7.2.5.2 Mechanism of the Gastrointestinal Disorder

Reflux which is the regurgitation of substances from the lower digestive tract into the higher organs is regarded as the first possible cause for the observed elevated incidence of overlaps in gastrointestinal illnesses. Numerous annulus muscles are known to operate as "gates" or one-way moving "check-points" along the gastrointestinal tract. These muscles include the orbicularis oris muscle, preventriculus, pylorus, Oddi sphincters, ileocecal valve, aperture of vermiform appendix, and the anus. When challenged by lower contents, all of the "gates" can withstand effacement and opening because they are fixed with smooth muscle or sphincter. If this isn't done, there will be instances of lower stomach acid reflux into upper digestive tracts (Ayazi et al. 2010). As a result of these sphincters' and smooth muscles' incapacity, one common mechanism depends on reflux. Irritable stimulation is an example of the second common pathophysiological mechanism. GI disorders frequently exhibit irritable comorbidity, which includes emotional irritability, rage, and depression. Most often, GI diseases are the cause of continuous emotional irritation in GI patients particularly in the form of anger and worry (Van Oudenhove et al. 2010). The relationships between GI function and symptom reporting and cognitive-affective processes have been partially explained by epidemiological, psychophysiological, and functional neuroimaging research. Studies have shown that antidepressant-using mental patients may be more susceptible to upper gastrointestinal hemorrhage. The halt of GI microcirculation is the third common mechanism responsible for the apparent overlaps of GI disorders. Catecholamine and dopamine fluctuations have been shown in function dyspepsia, IBS, and peptic ulcer.

7.2.5.3 Effects of the Gastrointestinal Disorders

Abnormal motility: Strong emotions or environmental stress can cause the esophagus, stomach, small intestine, and colon to move more quickly in healthy individuals. When compared to normal people, the FGIDs have an even higher motility response to stressors including psychological or physiological. Although there is some correlation between these motor responses and bowel symptoms, specifically diarrhea, constipation, and vomiting, it is inadequate to account for reports of persistent or recurrent stomach pain.

Intestinal hypersensitivity: Visceral hypersensitivity plays a role in describing why many functional GI illnesses (such as functional chest pain of suspected esophageal origin, epigastric pain syndrome, IBS, and FAPS) have poor associations between pain and GI motility. These patients may have an expanded area of somatic origin of visceral pain as well as a reduced pain threshold with balloon dilation of the colon (visceral hyperalgesia) or higher sensitivity even to normal intestinal function (e.g., allodynia). Patients with FGIDs may experience increased hypersensitivity, a condition known as sensitization or stimulus hyperalgesia. Such patients experience a longer-lasting and more severe rise in pain intensity following repeated colon balloon inflations compared to controls (Munakata et al. 1997).

Inflammation: Researchers have suggested that the development of symptoms may be influenced by increased inflammation in the neural plexi or intestinal mucosa. However, it wasn't until recently discovered that almost 50% of IBS patients had elevated activated mucosal inflammatory cells. This information seems to be related to other clinical observations that up to 25% of patients who present with an acute intestinal infection will go on to develop dyspeptic or IBS-like symptoms, and approximately one-third of patients with IBS or dyspepsia describe that their symptoms started after an acute enteric infection. These people often have higher levels of inflammatory cells and inflammatory cytokine expression in their mucosa (Gwee et al. 2003; Mearin et al. 2005; Chadwick et al. 2002).

7.2.5.4 Assessment of the Disease

All patients should benefit from a biopsychosocial approach to assessment, but those who are not responding to first-line medication therapy should receive particular attention. Assessment of the disease might be aided by collecting a psychosocial history.

7.2.5.5 Accessing History

In order to let the psychosocial events contributing to the illness to spontaneously unfold, the patient is encouraged to describe their history in his or her own unique style. The doctor's readiness to treat the psychological and biological components of illness should be conveyed through the questions. Maintaining eye contact, avoiding interruptions, and assuming a low control style are all suggested components of a patient-centered approach.

7.2.5.6 Medical Test

A positive diagnosis can be established with the application of symptom-based diagnostic criteria. Tests ought to be based on objective evidence such as blood in the stool, abnormal blood testing, etc. moreover, cost-effectiveness of the test as well as safety should also considered (Drossman et al. 1999).

7.2.6 Treatment of the Gastrointestinal Disorder

The treatment of the gastrointestinal disorder involves various approaches including therapeutic relationship, associating symptoms with psychosocial factors, and referral to a mental health professional.

7.2.6.1 Therapeutic Relationship

Therapeutic relationship can be achieved when the doctor asks the patient about their beliefs, worries, and expectations; listens and shows empathy when necessary; clears up any misunderstandings; works with the patient to negotiate a treatment plan.

7.2.6.2 Associating Symptoms with Psychosocial Factors

An evaluation of the relationship between gastrointestinal symptoms and psychosocial variables is facilitated by a daily log of the symptoms and the time of colon movements and the onset of menstruation, which are related to food adjustments, lifestyle modifications, or stress. Cognitive-behavioral techniques can be built upon this information. At the beginning of their condition, some patients find it difficult to acknowledge the importance of psychosocial factors; this is especially true for those who have experienced severe developmental trauma, such as sexual assault.

7.2.6.3 Referral to a Mental Health Professional

Referrals for consultation and treatment may be necessary for the (i) psychiatric disorders which require certain treatments such as antidepressants, cognitive-behavioral therapy, or other psychotherapies; (ii) a history of abuse that is discovered during consultation and may be impeding the patient's ability to adjust to the illness; (iii) serious impairment in daily functioning which calls for specific treatment to improve coping skills, and (iv) somatization, where a patient's symptoms are causing them to seek care from multiple specialists. The hypnotherapist employs gradual muscle relaxation together with relaxation suggestions to alleviate tension in the striated muscles and relax the smooth muscle of the gastrointestinal tract. Arousal reduction training, often known as relaxation training, demonstrates the patients how to combat the physiological effects of stress and anxiety.

7.2.6.4 Effects of Honey and Herbs on Gastrointestinal Disorder

Propolis, also known as "bee glue," is a waxy, resinous material. Bees generate honey and produce propolis by combining exudates from nearby plants, such as tree buds, sap flows, leaves, branches, and bark with their saliva and beeswax. Propolis is ultimately used by bees to cover fractures, shield themselves from predators and microbes, and provide thermal insulation for their colonies (Bankova et al. 2000). Propolis varies in color according to the kind of plant the bees utilized to gather the resinous materials. Propolis now frequently has been used to treat a variety of conditions, including gastrointestinal tract ailments such mucositis, colitis, gastritis, and peptic ulcers. A gastric ulcer is characterized as an injury to the stomach mucosa that results from an imbalance between the mucosa's resistance to the extremely acidic and proteolytic gastric juices and the luminal challenge they pose. Antisecretory medications, such as proton pump inhibitors (PPIs) and type-2 histamine receptor antagonists (H2-RAs) along with antibiotics for *H. pylori* infections, are the basis of treatment for stomach ulcers. Nevertheless, these medicinal products are usually linked to a host of unfavorable side effects including hypersensitivity, iron and vitamin B12 deficiency, arrhythmia, heightened vulnerability to pneumonia, impotence, gynecomastia, bone fractures, hematopoietic alterations, hypergastrinemia, and gastric cancer. Against this backdrop, natural products are viewed as appealing candidates for novel antiulcer therapies. Among these, propolis has been used in traditional medicine to treat gastric ulcers, which has stimulated research pertaining to the validation of its potential use as an antiulcer agent (Malfertheiner et al. 2009). Research on the gastroprotective properties of propolis ethanolic extract against ethanol-induced gastric ulcers in rats showed that administration of the extract inhibited the development of gastric ulcerations in a manner that is dose-dependent. Additionally, propolis extract scavenged and decreased the superoxide anion, as well as lipid peroxidation, based on both in vitro and in vivo investigations. Thus, the scientists deduced that the capacity of ethanol propolis extract to shield the stomach mucosa from oxidative stress is responsible for at least some of the gastric protective mechanism (Liu et al. 2002).

The major herbal therapies for functional gastrointestinal disorders include iberogast (STW-5) and peppermint oil. STW-5 is a formulation prepared from the extracts of medicinal herbs including *Iberis amara, Angelica radix, Carvi fructus Matricaria flos, Cardui mariae fructus, Mentha piperita, Chelidonii herba, Liquiritiae radix, and Melissa folium*. STW-5 was found to affect gastrointestinal motility and sensation by acting directly on muscles rather than exerting neural mechanisms of action. It caused the inhibition and relaxation of muscle activity dose dependently in the fundus region of the stomach. Peppermint oil obtained from the leaves of peppermint (*Mentha piperita*) exerts a spasmolytic effect on the smooth vasculature of the intestinal tract (Grigoleit and Grigoleit 2005; Kim et al. 2020).

7.2.7 Herbal Honey and Cardiovascular Disorders

A variety of illnesses affecting the heart and blood arteries are included in the category of cardiovascular disorders. They include peripheral arterial disease, aortic aneurysm and dissection, coronary heart disease, angina, stroke, rheumatic heart disease, congenital heart disease, deep vein thrombosis, and other, less prevalent, cardiovascular disorders. Myocardial infarction, a typical sign of cardiac disease, happens when there is insufficient blood flow from the coronary arteries to the myocardium, which results in hypoxia. Furthermore, ischemia may modify in situ metabolism by causing oxidative stress and inhibiting lipid metabolism, which might accumulate their harmful by-products in the heart (Greene et al. 2014). In

2003, cardiovascular diseases accounted for 16.7 million fatalities worldwide, or 29.2% of all deaths. It is the leading cause of mortality worldwide. Most Asian countries fall into the low- and middle-income category, accounting for 80% of all cardiovascular disease-related fatalities worldwide, despite the fact that heart attack mortality has decreased by more than 50% in several developed nations since the 1960s (Walsh 2004). In India, the percentage of urban residents with coronary disease has increased from 4% to 11% over the last 50 years. Patients under the age of 70 account for over half of all cardiovascular-related fatalities in India, but in the West, this number is just 22%. The potential effects of this trend on one of Asia's fastest-growing economies make it more concerning. Cardiovascular disease is the primary cause of mortality in India. In addition, it has been discovered that cardiovascular illness ranks third overall in terms of disease burden, after accidental and viral/parasitic disorders. Coronary heart disease killed 1.17 million Indians in 1990 and 1.59 million in 2000. Based on many surveys, a realistic estimate of 10% for individuals aged 35 and over in metropolitan areas has been investigated (Deb and Dasgupta 2008).

7.2.7.1 Coronary Heart Disease

The formation of atheromatous plaques leads to the development of coronary heart disease (CHD), also known as coronary artery disease (CAD) and atherosclerotic heart disease. A buildup of fatty material called plaque causes the artery lumen to constrict and obstruct blood flow. The process begins with the development of a "fatty streak." The formation of a fatty streak is caused by the subendothelial deposition of foam cells or lipid-loaded macrophages. The intima layer ruptures in response to a vascular insult, causing monocytes to go into the subendothelial region and develop into macrophages. Foam cells are created when these macrophages absorb oxidized low-density lipoprotein (LDL) particles. Activated T cells exclusively produce cytokines to support the pathogenic process. Smooth muscles are stimulated by growth hormones produced, which also cause them to absorb collagen and oxidized LDL particles, deposit them alongside active macrophages, and produce more foam cells. The result of this process is the development of subendothelial plaque (Shahjehan and Bhutta 2023). The primary cause of mortality in the majority of industrialized and developing nations is coronary heart disease or CHD. The major cause of the growing expense of healthcare is the clinical problems of congenital heart failure (CHD), which result in significant impairment. Non-modifiable and modifiable variables are the two main categories of etiologic factors. Non-modifiable risk factors for CHD include age, sex, and a history of cardiovascular disease in one's family and personally. The most modifiable risk factors include smoking cigarettes, having high blood pressure and having high cholesterol. Low levels of HDL cholesterol are thought to be a significant risk factor for CHD as well. In the treatment of congestive heart failure (CHD), other modifiable risk factors such as diabetes, obesity, and physical inactivity should be considered.

7.2.7.2 Heart Attack

Myocardial infarction (MI), sometimes referred to as a "heart attack" is brought on by a reduction in or interruption of blood flow to a section of the heart. Myocardial infarction can occur "silently" and go unnoticed, or it can be a devastating incident that results in hemodynamic decline and abrupt death. Oxygen is not available to the myocardium when coronary artery blockage occurs. Long-term oxygen supply deprivation of the heart can result in necrosis and death of heart cells. Patients may exhibit pressure or pain in the chest, which may extend to the jaw, arm, neck, or shoulder. Myocardial ischemia can be linked to increased biochemical markers such as cardiac troponins and abnormalities in the electrocardiogram in addition to the history and physical examination (Apple et al. 2017; Goodman et al. 2006). The modifiable risk factors for heart attack identified by the worldwide multi-center case-control research INTERHEART are consuming tobacco, abnormal blood apolipoprotein levels and lipid profiles (increased ApoB/ApoA1), high blood pressure, diabetes type I, and waist/hip ratio measurements indicating abdominal obesity (more than 0.90 for men and greater than 0.85 for women). Psychosocial variables include depression, losing sense of control, financial stress, and worldwide tension. Life events include divorce, losing one's job, disagreements with family members, not eating fruits or vegetables regularly, absence of exercise, and drinking alcohol. In the case of a heart attack, prompt action is essential for avoiding death.

7.2.7.3 Heart Failure

Heart failure (HF) is a clinical illness brought on by anatomical and functional flaws in the myocardium that limit blood ejection or ventricular filling. Reduced left ventricular myocardial function is the most frequent cause of heart failure (HF) nevertheless, pericardium, myocardium, endocardium, heart valves, or great vessels, either alone or in combination, can all malfunction and cause HF. A number of significant pathogenic factors can cause heart failure (HF) including increased hemodynamic overload, ischemia-related dysfunction, ventricular remodeling, excessive neuro-humoral stimulation, abnormal myocyte calcium cycling, excessive or insufficient extracellular matrix proliferation, accelerated apoptosis, and genetic mutations (Dassanayaka and Jones 2015). Depending on where the deficiency is located, heart failure can be categorized as primarily left ventricular, right ventricular or biventricular. Acute or chronic heart failure (HF) is determined by the time of onset. Heart failure with maintained ejection fraction (HFpEF) and heart failure with decreased ejection fraction (HFrEF) are the two main clinical kinds that are commonly distinguished by the functional state of the heart (Inamdar and Inamdar 2016). The left ventricular (LV) cavity volume is normally normal in patients with HFpEF, who are primarily female older people. However, the LV wall is thickened and rigid, which results in a high ratio of LV mass/end-diastolic volume (Ohtani et al. 2012). Acute coronary syndrome/ischemia, dysrhythmia/arrhythmias and alcohol intoxication or excess, thyroid conditions, pregnancy, and other iatrogenic conditions such as postoperative fluid replacement or the administration of steroids or non-steroidal anti-inflammatory drugs are among the common causes of heart failure. All of these conditions can either directly or indirectly contribute to the progression of the underlying disease.

7.2.7.4 Cardiac Arrhythmias

A cardiac arrhythmia may be described as any deviation from the normal heart rate or rhythm that is not supported by the body's physiological needs. Coordinated ion channel and transporter activity are necessary for the well-ordered transmission of electrical impulses across the heart and the maintenance of a regular cardiac rhythm. Interruptions to this well-organized mechanism result in heart arrhythmias which in certain cases can be fatal. The presence of structural cardiac disease resulting from myocardial infarction (owing to fibrotic scar development) or left ventricular dysfunction significantly increases the risk of commonly acquired arrhythmias (Kingma et al. 2023). Atrial fibrillation, a widespread cardiac arrhythmia primarily resulting from structural and electrical changes in atrial tissues is a supraventricular tachyarrhythmia brought on by uncoordinated atrial activation and atrial mechanical dysfunction. Although it can occur in the absence of underlying cardiac conditions, heart failure, ischemic heart disease, hypertension, and mitral valve disease are most commonly associated with it. The majority of cardiac arrhythmias are caused by structural myocardial illness, but they can also be brought on by a number of risk factors, including environmental, genetic, and altered epigenetic regulation (Ortmans et al. 2019). There is growing interest in identifying the gene(s) that cause inherited arrhythmogenesis. Several ion channel mutations that alter the cardiac action potential's configuration have been identified. It is also known that other mutations exist in genes that code for proteins involved in cytoskeletal construction, calcium handling, sodium transport, and cytokine signaling (Bermudez-Jimenez et al. 2018; Van der Heijden and Hassink 2013; Forleo et al. 2015). The function of inflammation in the development of atrial fibrillation is becoming more evident, despite the fact that its contribution to cardiac arrhythmias is often ignored. This is demonstrated by the acute rise in inflammatory proteins that have been detected.

7.2.7.5 Risk Factors for Cardiovascular Disorder

Genetic

Like majority of complicated disorders, cardiovascular disorder is influenced by a combination of hereditary and lifestyle variables. The idea that CAD risk is inherited has been validated by clinical data going back to the 1950s. This finding of a higher risk for CAD among close relatives was later supported by research including over 20,000 Swedish twins, which also indicated a heritability of about 50% for fatal CAD (Marenberg et al. 1994; Zdravkovic et al. 2002).

Physical Inactivity

One of the main risk factors for cardiovascular disorder (CVD) and stroke is physical inactivity. In addition to lowering premature mortality, physical activity also improves CVD risk factors including high blood pressure and cholesterol and lowers the chance of CVD-related illnesses like type 2 diabetes, heart attacks, stroke, and coronary heart disease. One to two million fatalities worldwide are attributed to physical inactivity, which is the fourth-leading risk factor for mortality. The risk of death was reduced for those over 40 who reported engaging in moderate-to-intense physical activity at least once a week than for those who did not.

Family History

Due to the hereditary component of cardiovascular disease, a family history of the illness is regarded as a risk factor. This usually happens when a person's first-degree relative experienced a significantly early onset of CVD. If the person's mother, sister, or father had cardiovascular disease before the age of 65, or if their father or brother had it before the age of 55, then this situation prevails. One's likelihood of acquiring high blood pressure (hypertension), high cholesterol, and type 2 diabetes can also be raised by a family history of these disorders. These factors can ultimately raise one's risk of cardiovascular disease.

Cholesterol

Numerous cardiovascular disorders are associated with elevated levels of low-density lipoprotein (LDL) cholesterol, sometimes referred to as "bad cholesterol." Proteins in the body transport fat or cholesterol, throughout the body. Excess LDL cholesterol can lead to difficulties by causing fatty substances to accumulate in the arterial walls. "Good cholesterol" is the term used to describe high-density lipoprotein (HDL) cholesterol. From all across the body, this cholesterol carries lipids and cholesterol to the liver, where they may be eliminated. Having a high level of HDL is typically advantageous since it can reduce the risk of heart disease and stroke, in contrast to LDL. Healthy eating, regular exercise, abstaining from tobacco, and consuming less alcohol can all assist to raise HDL cholesterol levels.

High Blood Pressure and Cardiovascular Disease

With the development of analytical methods, estimates of the high blood pressure contribution to the cause of CVD have steadily grown (Fuchs and Whelton 2019). For CHD and stroke, the early estimates of attributable risk were 25% and 50%, respectively (Kannel 1976). According to the risks determined by the Prospective Studies Collaboration, 49% of CHD cases and 62% of stroke cases were assessed to be linked to blood pressure equal to or greater than 115/75 mmHg (Lewington et al. 2002). An early cohort research examined the hazards associated with blood pressure, stratified by blood pressure levels, symptoms, abnormalities of the electrocardiogram, albuminuria/hematuria, and abnormalities of the optic fundus.

7.2.7.6 Diagnosis

Electrocardiogram (ECG or EKG). An ECG is a quick and painless test that records the electrical signals in the heart. It can tell if the heart is beating too fast or too slowly.
Holter monitoring. A Holter monitor is a portable ECG device that's worn for a day or more to record the heart's functioning during daily activities. This test can detect irregular heartbeats that aren't found during a regular ECG exam.
Echocardiogram. This non-invasive examination uses sound waves to create detailed images of the heart in motion. It shows how blood moves through the heart and heart valves. An echocardiogram can help determine if a valve is narrowed or leaking.
Exercise tests or stress tests. These tests often involve walking on a treadmill or riding a stationary bike while the heart is monitored. Exercise tests help reveal how the heart responds to physical activity and whether heart disease symptoms occur during exercise.
Cardiac catheterization. This test can show blockages in the heart arteries. A long, thin flexible tube (catheter) is inserted in a blood vessel, usually in the groin or wrist, and guided to the heart. Dye flows through the catheter to arteries in the heart. The dye helps the arteries show up more clearly on X-ray images taken during the test.
Heart (cardiac) CT scan. In a cardiac CT scan, a person lies on a table inside a doughnut-shaped machine. An X-ray tube inside the machine rotates around the person's body and collects images of person's heart and chest.
Heart (cardiac) magnetic resonance imaging (MRI) scan. A cardiac MRI uses a magnetic field and computer-generated radio waves to create detailed images of the heart.

7.2.7.7 Role of Herbs and Honey

The use of *Crataegus oxyacantha* as a pharmacologically active medicinal material dates back to many years. It is being used to treat angina, hypertension, arrhythmias, CHF, and other conditions. It has been used traditionally as a heart tonic (Miller 1998). According to reports, garlic (*Allium sativum*) inhibits the progression of cardiovascular disease and protects against cancer and other chronic illnesses linked to aging. According to scientific research, eating garlic can significantly reduce blood pressure, prevent atherosclerosis, lower serum cholesterol and triglycerides, inhibit platelet aggregation, and increase fibrinolytic activity (Chan et al. 2013). Guggulipid, an extract derived from the guggul (exudates of *Commiphora wightii*) lowers total blood cholesterol, triglycerides, and low-density cholesterol while raising HDL. In the indigenous medical system, fruits, bark, and leaves of *Terminalia arjuna* have been utilized to treat a variety of illnesses. It has been suggested that the bark powder of the plant possesses cardioprotective qualities. In his book "Ashtanga Hridayam," published around 1200 years ago, Vagbhata was the first to mention using powdered *T. arjuna* bark combined with milk to relieve heart-related chest discomfort. Green tea (*Camellia sinensis*) protects against coronary heart disease by a variety of pathways, including antioxidative, anti-inflammatory, antiproliferative, antiplatelet, and antithrombogenic effects (Velayutham et al. 2008; Islam 2012).

Regarding honey, several in vitro, in vivo, and clinical trial investigations have demonstrated that honey enhances the plasma lipid profile, which in turn influences heart disease risk factors (Alagwu et al. 2011). Numerous antioxidant enzymes, such as glucose oxidase and catalase as well as other antioxidant-containing substances such ascorbic acid, flavonoids, phenolic acids, derivatives of carotenoid pigments, organic acids, amino acids, and proteins have been shown to be present in honey. The primary preventive benefits of flavonoids in coronary heart disease are due to antithrombotic, anti-ischemic, antioxidant, and vasorelaxant properties. It is proposed that flavonoids reduce the incidence of coronary heart disease by three main mechanisms: (A) enhancing coronary vasodilatation, (B) reducing blood platelet clotting capacity, and (C) blocking LDL oxidation. Among the most well-known uses of honey is its ability to lower cholesterol in those with hyperlipidemia. For instance, cholesterol levels were dramatically lowered after receiving 75 g of honey diluted in 250 mL of water continuously for 15 days (Al-Waili 2004a, b). Another study examined the impact of honey on various health parameters such as fasting blood glucose (FBG), body weights, total cholesterol, high-density lipoprotein cholesterol (HDL-C), triacylglycerol, and C-reactive protein (CRP) in a group of 55 patients. The findings indicated that administering 70 g of honey orally for 30 days can lower LDL, triacylglycerols, and cholesterol in overweight patients (Yaghoobi et al. 2008). After 45 days of 3 g/kg/day Tualang honey treatment, rats with myocardial ischemia have shown a response. It has been observed that the disruption in the cardiac marker enzymes, aspartate transaminase (AST), lactate dehydrogenase (LDH), and creatine kinase-MB (CK-MB) has improved (Khalil et al. 2015).

7.2.8 Herbal Honey and Reproductive Disorders

Human reproductive abnormalities can have a major impact on fertility as well as general health of the people. Numerous reproductive health problems can affect both men and women and can be caused by genetic, environmental, hormonal, or lifestyle factors. It has been found that reproductive diseases including uterine fibroids, endometriosis, adenomyosis, and polycystic ovarian syndrome (PCOS) may negatively affect pregnancy. With the passage of time, unfavorable patterns have been observed in male reproductive health such as a rise in the prevalence of testicular cancer, weak and very likely deteriorating semen quality, decrease in male fertility, as well as an increase in cases of cryptorchidism. The "estrogen hypothesis" was proposed in 1993 as a result of evidence of a higher incidence of hypospadias and cryptorchidism as well as a well-documented rise in testicular cancer in a number of nations (Sharpe and Skakkebaek 1993). It was postulated that elevated levels of estrogen, particularly in the fetal stage, might potentially disrupt the endocrine regulation of the male fetal urogenital organs (Storgaard et al. 2006). Obesity and overweight pose a serious danger to the health of populations in a growing number of nations globally. Reproductive endocrinology has observed significant correlations between excess body fat, especially abdominal obesity, and irregular menstrual cycles, decreased spontaneous and induced fertility, and increased risk of miscarriage and hormone-sensitive carcinomas.

7.2.8.1 Polycystic Ovary Syndrome

PCOS is the condition that affects the human ovary and causes ovarian-dependent infertility. Polycystic ovarian syndrome is thought to be the most prevalent endocrine disorder in women of reproductive age. About half of women suffering with PCOS are fat or overweight and have decreased insulin sensitivity (Norman et al. 2004). It has been found that PCOS impairs fertility and the quality of pregnancy outcomes. Newborns to mothers with PCOS are more likely to experience neonatal complications such as preterm birth, perinatal death, and more admissions to critical care units. A wide range of clinical symptoms, such as polycystic ovarian morphology, menstrual dysfunction/oligo-anovulation, and hyperandrogenism are seen in PCOS patients. Infertility, abnormal uterine hemorrhage, metabolic problems (dyslipidemia, obesity, type 2 diabetes mellitus, and cardiovascular disease are among the conditions that women with PCOS are more likely to have.

7.2.8.2 Endometriosis

One of the main causes of infertility and persistent pelvic pain in women is endometriosis, which is defined by ectopic endometrium, or the presence of endometrial glands and stroma outside the uterus. The only tissue that renews itself every month is the endometrium. Although estimates of the prevalence of disease vary, most research indicates that between 10% and 15% of women in reproductive age have endometriosis. Infertility and pelvic pain are associated with a substantially greater incidence (between 35% and 50%) in women. It is debatable whether endometriosis patients who become pregnant naturally or with ART have a higher risk of obstetric problems. Specifically, women who have endometriosis during their first pregnancy are more likely to experience gestational diabetes mellitus (GDM), premature preterm rupture of membranes (pPROMs), pre-term birth, and lengthier hospital stays for both the mother and the newborn (Conti et al. 2014).

According to the Samson idea, often known as "retrograde menstruation," menstrual blood is displaced through the fallopian tubes and enters the peritoneal cavity, causing the foci of endometriosis. Data from published studies show that retrograde menstrual blood flow occurs in 80% of women with open fallopian tubes, whereas endometriosis only affects a small percentage of these women. This shows that there may be other variables influencing endometrial cells' ability to survive and implant in the peritoneal cavity. The onset, development, and progression of endometriosis are influenced by increased release of pro-inflammatory cytokines and decreased production of anti-inflammatory substances from stromal, epithelial, smooth, or immune cells (Asghari et al. 2018). New artery development is facilitated by vascular endothelial growth factor or VEGF. Elevated levels of VEGF in peritoneal fluid were observed in endometriosis-affected women, and this level was found to correlate with the disease's phases.

7.2.8.3 Uterine Fibroids

As the most prevalent benign tumors of the female reproductive system, uterine fibroids (leiomyomas) afflict between 30% and 70% of women of reproductive age. As a result, they account for 0.1–12.5% of all pregnancies (Cooper and Okolo 2005). Fibroids are linked to complications during pregnancy including antepartum, intrapartum, and postpartum issues. They are the primary cause of hysterectomies in women of reproductive age. In the premenopausal years, the chance of developing leiomyoma tumors increases with age; however, tumors usually shrink and/or become asymptomatic with the start of menopause. Preterm birth is the most frequent reason why pregnant women with fibroids experience newborn morbidity (Lai et al. 2012).

Growing older is a substantial risk factor for uterine fibroids, particularly for premenopausal women and those over 40. For example, 80% of African American women over 50 have uterine fibroids, compared to 60% of those between the ages of 35 and 49. Forty percent of White women under 35% and 70% of those over 50 had uterine fibroids (Baird et al. 2003). Numerous investigations have revealed obesity to be a major risk factor for the development of uterine fibroids. It has been linked to the adipose tissues' metabolic processes. A variety of cytokines and growth factors, including those involved in the regulation of immune and inflammatory responses are produced and released by adipose tissues (Ellulu et al. 2017). Arterial hypertension and uterine fibroids are directly correlated. Uterine fibroids are five times as common in women with hypertension (Takeda et al. 2008). The development of uterine fibroids is influenced by lifestyle variables, including nutrition and degree of physical activity.

7.2.8.4 Cryptorchidism

One of the most prevalent congenital abnormalities of the male genitalia is cryptorchidism, also known as undescended testis [UDT], which affects about 1% of males. Congenital cryptorchidism is the term for the condition when one or both testes are born without normally falling to the bottom of the scrotum. By week seven of pregnancy, the SRY gene is situated on the Y chromosome's short arm. The mullerian-inhibiting substance (MIS), which is produced by Sertoli cells under the influence of the SRY gene, causes the mullerian duct cells to invert. The formation of normal male internal genitalia occurs during the eighth week of gestation and is noticeable by weeks 10–13. This is because MIS generated by fetal testicular Sertoli cells and testicular androgens produced by fetal testicular Leydig cells are responsible for this development (Braga and Lorenzo 2017). Cryptorchidism can arise from a disturbance at any stage of testicular descent caused by genetic, hormonal, anatomical, environmental, or social reasons. Testis cancer, infertility, and hypospadias are all at risk due to cryptorchidism.

7.2.8.5 Testicular Germ Cell Cancer

One of the most prevalent cancers among males between the ages of 15 and 45 is testicular cancer. Multiple genetic and environmental variables contribute to the multifactorial etiology. It accounts for 5% of urological malignancies and 1% of tumors in men (Park et al. 2018). The prognosis is favorable with >90% cure rate and >95% five-year survival rate with good care (Smith et al. 2018).

7.2.8.6 Hypospadias

The congenital abnormality known as hypospadias affects the penile region, causing the urethra to open somewhere on the ventral side of the penis rather than the tip. Surgery is not always necessary to address this birth abnormality. This abnormality may also be hidden by physiological phimosis in the infant, showing up only once the foreskin can be readily retracted (Boisen et al. 2005).

7.2.8.7 Cause of Reproductive Disorders

Men can have reproductive abnormalities for a variety of reasons, and these causes frequently include a mix of genetic, environmental, lifestyle, and health-related variables. The following are some common factors linked to both male and female reproductive disorders:

Genetic

Over the course of the last 30 years, a number of genetic disorders pertaining to hormone production and receptors have been identified. These flaws are frequently linked to cryptorchidism as part of a syndrome, but they are extremely uncommon in individuals with isolated cryptorchidism that is, without additional genital abnormalities (Massart and Saggese 2010). Certain gene abnormalities, such as those affecting the AMH gene or its receptor (AMHR2) in persistent Mullerian Duct syndrome, may produce solitary cryptorchidism by physically impairing testicular descent (Abduljabbar et al. 2012; Josso et al. 1993). Hypospadias is caused by a number of genetic abnormalities that are known to be associated with problems of androgen receptor function, testosterone synthesis, testosterone conversion to dihydrotestosterone, or testicular differentiation (Kalfa et al. 2009). Genes encoding signaling components linked to steroidogenesis, steroid hormone action, gonadotrophin action and regulation, insulin action and secretion, energy metabolism, and chronic inflammation are most commonly implicated in the pathogenesis of PCO (Khan et al. 2019).

Hormonal

Hormonal regulation governs the descent of the testicles. Insulin-like peptide 3 (INSL3) and testosterone are the main regulatory hormones, and the testis secretes both of these substances from Leydig cells. Leydig cell development and hormone release are stimulated by pituitary luteinizing hormone (LH). The testes stay partially descended when there is a reduction in these hormones or a malfunction in their receptors. Androgens are the primary cause of male masculinization in the early stages of fetal development. Dihydrotestosterone, which is generated locally from testosterone by 5-reductase, controls penile growth (Skakkebaek et al. 2016).

Lifestyle

Lifestyle variables including drinking alcohol and smoking are important risk factors for cryptorchidism; however, there is disagreement about their relative importance. Evidence suggests that there is a higher chance of having a bilaterally cryptorchid boy if a pregnant woman smokes heavily around 10 cigarettes a day (Thorup et al. 2006).

Role of Herbs and Honey in the Treatment of Reproductive Disorders

Numerous investigations have demonstrated the direct estrogenic effects of honey on the female reproductive system. Three-month-old Sprague Dawley rats treated with Tualang honey at varying dosages demonstrated enhanced epithelial thickness, vaginal and uterine weights, and antioxidant capacity (Zaid et al.2010). The estrogenic activity of honey supplementation was attributed to flavonoids and phenolic acids with the result that menopausal symptoms were somewhat alleviated (Mosavat et al. 2014). In a different study, 1.2 g/kg/day of honey was given orally to mitigate the effects of smoke-induced reproductive damage. Consequently, higher fertility and mating rates are the outcome. A preliminary clinical investigation was carried out on postmenopausal females between the ages of 45 and 65. After undergoing a 6-month follow-up, 200 postmenopausal healthy women were administered 20 g of Tualang honey daily. The findings demonstrated that, after 6 months, there were notable differences in the two groups' diastolic blood pressure, serum total cholesterol, and LDL levels (Waykar and Alqadhi 2016).

Honey has been shown to preserve sexual behavior in rats exposed to cigarette smoke (CS) and shield rat testes from oxidative damage (Mohamed et al. 2011; Mohamed et al. 2013). As a spermiogenesis booster and in semen extenders, it also maintained the quality of sperm. Supplementing normal rats with honey enhanced their spermatogenesis. It was found in another study that 10% of honey (1 mL of honey and 9 mL of IVF culture medium) at doses of 1.2 and 1.8 g/kg bw increased the diameters of seminiferous tubules, increased testosterone, FSH, and LH hormones, and improved sperm motility (Hadi 2017; Mohamed et al. 2013).

Walnut (*Juglans regia*) oil may raise plasma testosterone levels by altering the pituitary-testicular axis having stimulating effects on the male reproductive system (Bostani et al. 2014). The pituitary-gonadal axis hormones (FSH, LH, and testosterone) were shown to rise in an extract of parsley (*Petroselinum crispum*) leaves because of the presence of antioxidant chemicals (Bastampoor et al. 2014). In adult female mice, it has been demonstrated that the palm pollen aqueous extract increases the quantity of sexual hormones and the numbers of secondary and antral follicles (Hosseini et al. 2014). Scientifically known as, cinnamon is a member of the Lauraceae family, which also includes antihypertensive drugs (Hosseini et al. 2014). Among the several medicinal benefits of this herb is an increase in libido. Cinnamon (*Cinnamomum verum*) extract has beneficial effects on the male reproductive system and pituitary-gonad axis hormonal alterations since it increases sperm count and FSH hormone output (Modaresi et al. 2009). *Ruta graveolens* has been used to induce abortion in women and sexual dysfunction in males. The aqueous extract of this plant reduces the activity of reproductive system and may be helpful as a component of birth control (Ahmadi et al. 2007). The use of ginseng (*Panax ginseng*) extract thickens the uterine wall and endometrial layer while decreasing the number of atretic follicles and increasing the number of primary, secondary, and graafian follicles. Thus, uterine tissue and oogenesis benefit from the use of red ginseng water extract (Zohrevand Asl et al. 2017). In animals with polycystic ovarian dysfunction, there were more atretic follicles, higher levels of estrogen and testosterone, and fewer other follicles and FSH. Additionally, the use of metformin chemical and extract of ginger and ginseng was shown to decrease the number of atretic follicles, FSH, and other follicles in these animals, as well as lower levels of testosterone and estrogen (Foroozandeh and Hosseini 2017; Shabani and Hosseini 2017). *Aloe vera* hydro-alcoholic extract raised progesterone hormone concentration and reduced estrogen levels in rats, respectively. It could be used to treat PCOS and infertility related issues (Hemayatkhah-Jahromi and Rahmanian Koushkaki 2016).

Chapter 8
Critical Analysis and Future Perspectives

Honey was one of man's first foods that helped him get established in this world. It has been valued and given due importance by people all over the globe since time immemorial. It has been defined as aromatic, viscid sweet material derived from the nectar of plants by honeybees, modified by them into denser liquid and finally stored in combs. It is chiefly composed of sugars besides acids, enzymes, vitamins, ash, moisture, pollen grains, etc.

Honey is a natural, unprocessed food and is the only sweet available in commercial quantities. It possesses taste and aroma depending upon the type of flora from where nectar is obtained. Along with nectar, many plant components like flavonoids, polyphenols, etc. also reach inside the beehive, due to which it may also be called herbal honey. At one time, honeybees visit only one floral source until it gets dried. Since at one time blossoms of one floral source predominate, it is possible to produce herbal honey of a specific kind if mustard is rich in area, then it will be mustard honey, if sunflower bloom is there then it is named sunflower honey and so on (Table 8.1).

Honey is prized for its flavor and therapeutic properties that can be enjoyed by tasting it. Most of the components present in honey get assimilated into the digestive system and are a ready source of energy. Under natural conditions, herbal honey can directly be obtained by placing the bee colonies in the herb-rich area. Artificial herbal honey may be obtained by infusing extracts of herbs in honey by artificial means in the laboratory as described in Chap. 5. Honey alone is considered to be miraculous due to its immense pharmacological properties and a wide number of benefits to mankind, however, when infused with herbs its potential increases many folds (Refer to Chap. 7). Honey supplemented with herbs may possess enhanced therapeutic activities that can contribute to the management of various communicable and non-communicable diseases. Many researchers have made efforts to infuse herbal extract into honey and tested infused honey against various diseases (Refer to Chap. 5). Magnificent results have been obtained in the majority of cases where patients benefitted, however, more attention and effort are required in this field so that it can be made commercially available for the benefit of mankind.

Table 8.1 Types of herbal honey depending upon the type of flora

S. no.	Honey type	Dominant floral source
1	Mustard honey	Mustard
2	Apple honey	Apple
3	Litchi honey	Litchi
4	Chestnut honey	Chestnut
5	Berry honey	Berry
6	Eucalyptus honey	Eucalyptus
7	Citrus honey	Lemon/Orange
8	Acacia honey	Kikar/Phalai/Khair
9	Mulberry honey	Mulberry
10	Plectranthus honey	Plectranthus
11	Soapnut honey	Sapindus
12	Shisham honey	Dalbergia
13	White willow honey	Salix
14	Cashew honey	Anacardium
15	Manuka honey	Manuka
16	Ber honey	Ziziphus
17	Saffron honey	Crocus
18	Castor honey	Ricinus
19	Jamun honey	Syzygium
20	Clove honey	Clove

The readers will benefit after reading various chapters of this book. Efforts have been made to describe various mechanisms of herb-infused honey using flow diagrams, pictorial presentations, etc. Chap. 1 illustrates the comprehensive details about the general description, history, types, composition, preservation, and storage of honey. Chapter 2 highlights the biological activities possessed by honey and its therapeutic significance in combating infection. All the properties have been described using suitable examples wherever required. Chapter 3 includes the therapeutic potential of valuable herbal resources in Global, Indian, and Himalayan contexts. Original photographs of different herbs have been presented for better understanding of readers. Chapter 4 provides an in-depth understanding of the relationship between plants and honeybees. Flora of bee interest has been described along with photographs. In Chap. 5, methods used for infusing herbs in honey have been compiled and described by citing the work of previous authors. Also, the purpose of producing herb-infused honey has been described. Chapter 6 highlights the factors that make herbs more valuable for mankind after infusion with honey. Chapter 7 is the main attraction of this book which describes the remedial aspects of herbal honey. The chapter has been divided into two sections: communicable and non-communicable diseases. The mechanisms have been drawn to make the readers clearer about the action of herbal honey in the mitigation of various diseases. Last and current Chap. 8 describes challenges and future perspectives of herbal honey for the betterment of human health and mankind.

8.1 Challenges

There are certain challenges in the field of formulating herbal honey, few of them are as follows:

(i) Availability of pure honey
(ii) Selection of herb
(iii) Selection of infusion method
(iv) Competition with allopathic system
(v) Suitability to patients
(vi) Commercialization

(i) Availability of Pure Honey
   The biggest hurdle in the production of herbal honey is the availability of pure honey. A lot of cases of honey adulteration are being reported in different parts of the world. If impure honey is infused with herbs, it may produce negative impacts on health and hygiene.
(ii) Selection of Herb
   The selection of herbs for its infusion with honey is yet another challenge. Herbs have certain properties that produce specific effects on diseases. Also, herbs consist of polyphenols, alkaloids, free radicals, etc. that may have varying effects on human health. Therefore, the selection of the right herb is important to produce good quality herbal honey.
(iii) Selection of Infusion Method
   Herb-infused honey can be prepared using different infusion methods. It is important to decide the exact and valid method of infusion so that herb extract can be fully infused in honey and the best decoction can be prepared.
(iv) Competition with Allopathic System
   The current world has entered an era where people have become habitual toward an allopathic system of medicine for resolving health-related issues. Everybody wants quick relief for diseases/any health issue. However, in many cases, there are many associated health issues and the problem may resurge which is a big problem. Hence, herbal honey formulations have strong competition with allopathic medicines.
(v) Suitability to Patients
   Another issue with the herbal honey formulation is its suitability for patients suffering from particular diseases/disorders. Some patients may have side effects from a particular formulation, whereas others may benefit from the same. Therefore, the suitability of herbal honey formulation should be checked.
(vi) Commercialization
   The sale and commercialization of herbal honey formulations are yet other issues. There should be some good brands for popularizing and selling these formulations.

8.2 Future Perspectives

Herbal honey has many perspectives in research and medical field. Some future perspectives are discussed as follows:

(i) Boon in health science
(ii) Revolution in the honey industry
(iii) Promotion of herbal plants
(iv) Research opportunities

(i) Boon in Health Science
Herbal honey may bring a boon in the health/medical field due to the least side effects. People are facing lots of issues while using allopathic medicines. Many times, associated symptoms may appear that further aggravate the health conditions. Use of herbal honey treatments may solve this issue.

(ii) Revolution in the Honey Industry
The honey industry has not been promoted the way it should be. Beekeepers or honey dealers are still getting meager value for honey. Herbal honey formulations may raise the demand for honey which will increase its commercial value and the economic standard of beekeepers may be raised.

(iii) Promotion of Herbs
Herbal plants have been used by the human race for the cure of diseases since time immemorial. Still, herbs have not gained that much importance. The people growing herbs are still not getting enough value. So, using herbs in the honey industry may promote herbs and help to increase the economic value of herbs.

(iv) Research Opportunities
The herbal honey and its use in the health/medical field will certainly promote research in the field.

8.3 Recommendations

Following recommendations may be made toward herbal honey and its biomedical perspectives:

(i) More research is required in the field of herbal honey to strengthen its benefits for human health.
(ii) Emphasis should be laid upon validating infusion methods for herbs and honey.
(iii) Industry academia relation should be strengthened for popularizing honey industry.
(iv) People should be made more aware about herbal honey and its importance in human health.

Bibliography

Abdellah F, Abderrahim LA (2014) Honey for gastrointestinal disorders. In: Honey in traditional and modern medicine. CRC Press, pp 159–186. https://doi.org/10.1201/b15608-9

Abduljabbar M, Taheini K, Picard JY, Cate RL, Josso N (2012) Mutations of the AMH type II receptor in two extended families with persistent Mullerian duct syndrome: lack of phenotype/genotype correlation. Horm Res Paediatr 77:291–297

Abdulrhman M, Shatla R, Mohamed S (2016) The effects of honey supplementation on Egyptian children with hepatitis A: a randomized double blinded placebo controlled pilot study. J Apither 1(1):23. https://doi.org/10.5455/ja.20160702011113

Abuelgasim H, Albury C, Lee J (2021) Effectiveness of honey for symptomatic relief in upper respiratory tract infections: a systematic review and meta-analysis. BMJ Evid-Based Med 26(2):57–64

Adhikari PP, Paul SB (2018) History of Indian traditional medicine: a medical inheritance. Asian J Pharm Clin Res 11(1):421–426

Adnan F, Sadiq M, Jehangir A (2011) Anti-hyperlipidemic effect of acacia honey (desi kikar) in cholesterol-diet induced hyperlipidemia in rats. Biomedica 27(13):62–67

Afroz R, Tanvir E, Zheng W, Little P (2016) Molecular pharmacology of honey. J Clin Exp Pharmacol 6:212. https://doi.org/10.4172/2161-1459.1000212

Aggarwal BB, Kunnumakkara AB, Harikumar KB, Tharakan ST, Sung B, Anand P (2008) Potential of spice-derived phytochemicals for cancer prevention. Planta Med 74(13):1560–1569

Ahmad I, Beg AZ (2001) Antimicrobial and phytochemical studies on 45 Indian medicinal plants against multi-drug resistant human pathogens. J Ethnopharmacol 74(2):113–123

Ahmad A, Riaz S, Farooq R, Ahmed M, Hussain N (2023) *Alpinia officinarum* (galangal): a beneficial plant. J Med Public Health 4(1):1057

Ahmadi A, Nasiri Nejad F, Parivar K (2007) Effect of aqueous extract of the aerial part of the *Ruta Graveolens* on the spermatogenesis of immature Balb/C Mice. RJMS 14(56):13–20

Ahmed S, Sulaiman SA, Baig AA, Ibrahim M, Liaqat S, Fatima S et al (2018) Honey as a potential natural antioxidant medicine: an insight into its molecular mechanisms of action. Oxid Med Cell Longev 2018:8367846. https://doi.org/10.1155/2018/8367846

Akbik D, Ghadiri M, Chrzanowski W, Rohanizadeh R (2014) Curcumin as a wound healing agent. Life Sci 116(1):1–7. https://doi.org/10.1016/j.lfs.2014.08.016

Akkoca M, Kocaay AF, Tokgoz S, Er S, Duman B, Ayaz T et al (2022) Psychiatric symptoms, aggression, and sexual dysfunction among patients with benign anal conditions. Am Surg. https://doi.org/10.1177/00031348221074225

Akram M, Thiruvengadam M, Zainab R, Daniyal M, Bankole MM, Rebezov M, Shariati MA, Okuskhanova E (2022) Herbal medicine for the management of laxative activity. Curr Pharm Biotechnol 23(10):1269–1283

Alagwu E, Okwara J, Nneli R, Osim E (2011) Effect of honey intake on serum cholesterol, triglycerides and lipoprotein levels in albino rats and potential benefits on risks of coronary heart disease. Niger J Physiol Sci 26:161–165

Alam F, Islam M, Gan SH, Khalil M (2014) Honey: a potential therapeutic agent for managing diabetic wounds. Evid Based Complement Alternat Med 2014:169130. https://doi.org/10.1155/2014/169130

Alandejani T, Marsan J, Ferris W et al (2009) Effectiveness of honey on Staphylococcus aureus and Pseudomonas aeruginosa biofilms. Otolaryngol Head Neck Surg 141(1):114–118

Al-Hatamleh MAI, Hatmal MM, Sattar K, Ahmad S, Mustafa MZ, Bittencourt MDC, Mohamud R (2020) Antiviral and immunomodulatory effects of phytochemicals from honey against COVID-19: potential mechanisms of action and future directions. Molecules 25(21):5017. https://doi.org/10.3390/molecules25215017

Almasaudi S (2021) The antibacterial activities of honey. Saudi J Biol Sci 28(4):2188–2196

Al-Waili NS (2004a) Topical honey application vs. acyclovir for the treatment of recurrent herpes simplex lesions. Med Sci Monit 10(8):MT94–MT98

Al-Waili NS (2004b) Natural honey lowers plasma glucose, C-reactive protein, homocysteine, and blood lipids in healthy, diabetic, and hyperlipidemic subjects: comparison with dextrose and sucrose. J Med Food 7:100–107

Al-Waili NS, Saloom KS, Al-Waili TN, Al-Waili AN (2006) The safety and efficacy of a mixture of honey, olive oil, and beeswax for the management of hemorrhoids and anal fissure: a pilot study. Sci World J 6:1998–2005. https://doi.org/10.1100/tsw.2006.333

Al-Waili N, Salom K, Al-Ghamdi A, Ansari MJ, Al-Waili A, Al-Waili T (2013) Honey and cardiovascular risk factors, in normal individuals and in patients with diabetes mellitus or dyslipidemia. J Med Food 16(12):1063–1078

Alzergy AA, Elgharbawy SM, Mahmoud GS, Mahmoud MR (2015) Role of *Capparis spinosa* in ameliorating trichloroacetic acid induced toxicity in liver of Swiss albino mice. Life Sci J 12:26–39

Amal MNA, Koh CB, Nurliyana M, Suhaiba M, Nor-Amalina Z, Santha S, Diyana-Nadhirah KP, Yusof MT, Ina-Salwany MY, Zamri-Saad M (2018) A case of natural co-infection of Tilapia Lake Virus and Aeromonas veronii in a Malaysian red hybrid tilapia (Oreochromis niloticus× O. mossambicus) farm experiencing high mortality. Aquaculture 485:12–16

Ames BN, Shigenaga MK, Hagen TM (1993) Oxidants, antioxidants, and the degenerative diseases of aging. Proc Natl Acad Sci U S A 90(17):7915–7922. https://doi.org/10.1073/pnas.90.17.7915

Andreu V, Mendoza G, Arruebo M, Irusta S (2015) Smart dressings based on nanostructured fibers containing natural origin antimicrobial, anti-inflammatory, and regenerative compounds. Materials 8(8):5154–5193

Aparna S, Srirangarajan S, Malgi V, Setlur KP, Shashidhar R, Setty S, Thakur S (2012) A comparative evaluation of the antibacterial efficacy of honey in vitro and antiplaque efficacy in a 4-day plaque regrowth model in vivo: preliminary results. J Periodontol 83(9):1116–1121. https://doi.org/10.1902/jop.2012.110461

Apple FS, Sandoval Y, Jaffe AS, Ordonez-Llanos J, IFCC Task Force on Clinical Applications of Cardiac Bio-Markers (2017) Cardiac troponin assays: guide to understanding analytical characteristics and their impact on clinical care. Clin Chem 63(1):73–81

Ara I, Maqbool M, Zehravi M, Gani I (2020) Herbs boosting immunity in Covid-19: an overview. Adv J Chem B 3(3):289–294. https://doi.org/10.22034/ajcb.2021.299322.1091

Arduino PG, Porter SR (2008) Herpes Simplex Virus Type 1 infection: overview on relevant clinico-pathological features. J Oral Pathol Med 37(2):107–121. https://doi.org/10.1111/j.1600-0714.2007.00586.x

Ariyamuthu R, Albert VR, Je S (2022) An overview of food preservation using conventional and modern methods. J Food Nutr Sci 10(3):70–79

Arunachalam KD, Subhashini S, Annamalai SK (2012) Wound healing and antigenotoxic activities of *Aegle marmelos* with relation to its antioxidant properties. J Pharm Res 5(3):1492–1502

Asghari S, Valizadeh A, Aghebati-Maleki L, Nouri M, Yousefi M (2018) Endometriosis: perspective, lights, and shadows of etiology. Biomed Pharmacother 106:163–174

Aslanova M, Ali R, Zito PM (2023) Herpetic Gingivostomatitis. In: StatPearls. StatPearls Publishing, Treasure Island, FL. PMID: 30252324

Attia W, Gabry M, El-Shaikh K, Othman G (2008) The anti-tumor effect of bee honey in Ehrlich ascite tumor model of mice is coincided with stimulation of the immune cells. Egypt J Immunol 15:169–183

Atwa ADA, Abu Shahba RY, Mostafa M, Hashem MI (2014) Effect of honey in preventing gingivitis and dental caries in patients undergoing orthodontic treatment. Saudi Dent J 26(3):108–114. https://doi.org/10.1016/j.sdentj.2014.03.001

Aurongzeb M, Azim MK (2011) Antimicrobial properties of natural honey: a review of literature. Pak J Biochem Mol Biol 44(3):118–124

Awad OGAN, Hamad AMH (2018) Honey can help in herpes simplex gingivostomatitis in children: prospective randomized double blind placebo controlled clinical trial. Am J Otolaryngol 39(6):759–763. https://doi.org/10.1016/j.amjoto.2018.09.007

Ayazi S, Tamhankar A, DeMeester SR, Zehetner J, Wu C, Lipham JC, Hagen JA, DeMeester TR (2010) The impact of gastric distension on the lower esophageal sphincter and its exposure to acid gastric juice. Ann Surg 252(1):57–62. https://doi.org/10.1097/SLA.0b013e3181e3e41

Baird DD, Dunson DB, Hill MC, Cousins D, Schectman JM (2003) High cumulative incidence of uterine leiomyoma in black and white women: ultrasound evidence. Am J Obstet Gynecol 188(1):100–107

Balestrieri ML, Dicitore A, Benevento R, Di Maio M, Santoriello A, Canonico S, Giordano A, Stiuso P (2012) Interplay between membrane lipid peroxidation, transglutaminase activity, and Cyclooxygenase 2 expression in the tissue adjoining to breast cancer. J Cell Physiol 227:1577–1582. https://doi.org/10.1002/jcp.22874

Banker D (2003) Viral hepatitis (part-I). Indian J Med Sci 57(8):363–368. https://tspace.library.utoronto.ca/html/1807/23885/ms03016.html

Bankova VS, de Castro SL, Marcucci MC (2000) Propolis: recent advances in chemistry and plant origin. Apidologie 31(1):3–15

Basma AA, Zakaria Z, Latha LY, Sasidharan S (2011) Antioxidant activity and phytochemical screening of the methanol extracts of *Euphorbia hirta* L. Asian Pac J Trop Med 4(5):386–390

Bastampoor F, Sadeghi H, Hosseini SE (2014) The *Petroselinum crispum* L. hydroalcoholic extract effects on pituitary gonad axis in adult rats. Armaghane Danesh J 19(4):305–313

Battino M, Forbes-Hernández TY, Gasparrini M, Afrin S, Cianciosi D, Zhang J et al (2019) Relevance of functional foods in the Mediterranean diet: the role of olive oil, berries and honey in the prevention of cancer and cardiovascular diseases. Crit Rev Food Sci Nutr 59(6):893–920

Battino M, Giampieri F, Cianciosi D, Ansary J, Chen X, Zhang D et al (2021) The roles of strawberry and honey phytochemicals on human health: a possible clue on the molecular mechanisms involved in the prevention of oxidative stress and inflammation. Phytomedicine 86:153170

Becerril-Sánchez AL, Quintero-Salazar B, Dublán-García O, Escalona-Buendía HB (2021) Phenolic compounds in honey and their relationship with antioxidant activity, botanical origin, and color. Antioxidants 10(11):1700

Bellamy C, Jenkins SJ, McSorley HJ, Dorward DA, Kendall TJ (2020) Inflammation and immunology. In: Muir's textbook of pathology. CRC Press, pp 37–58

Bermudez-Jimenez FJ, Carriel V, Brodehl A, Alaminos M, Campos A, Schirmer I, Milting H, Abril B, Alvarez M, Lopez-Fernandez S et al (2018) Novel desmin mutation p.Glu401Asp impairs filament formation, disrupts cell membrane integrity, and causes severe arrhythmogenic left ventricular cardiomyopathy/dysplasia. Circulation 137:1595–1610

Bharucha AE, Wald AM (2010) Anorectal disorders. Am J Gastroenterol 105(4):786. https://doi.org/10.1038/ajg.2010.70

Bhattarai, B. (2021). Effect of honey proportion and pH on the sensory quality of mead. (Doctoral dissertation, Department of Food Technology Central Campus of Technology Institute of Science and Technology Tribhuvan University, Nepal 2019)

Biluca FC, da Silva B, Caon T, Mohr ETB, Vieira GN, Gonzaga LV et al (2020) Investigation of phenolic compounds, antioxidant and anti-inflammatory activities in stingless bee honey (Meliponinae). Food Res Int 129:108756

Blackadar CB (2016) Historical review of the cause of cancer. World J Clin Oncol 7(1):54–86

Boisen KA, Chellakooty M, Schmidt IM, Kai CM, Damgaard IN, Suomi AM, Toppari J, Skakkebæk NE, Main KM (2005) Hypospadias in a cohort of 1072 Danish newborn boys: prevalence and relationship to placental weight, anthropometrical measurements at birth, and reproductive hormone levels at 3 months of age. J Clin Endocrinol Metab 90(7):4041–4046

Borlinghaus J, Albrecht F, Gruhlke MC, Nwachukwu ID, Slusarenko AJ (2014) Allicin: chemistry and biological properties. Molecules 19(8):12591–12618

Bostani M, Aqababa H, Hosseini SE, Changizi Ashtiyani S (2014) A Study on the effects of walnut oil on plasma levels of testosterone pre and post puberty in male rats. Am J Ethnomed 1(4):266–275

Braga LH, Lorenzo AJ (2017) Cryptorchidism: a practical review for all community healthcare providers. Can Urol Assoc J 11(1–2 Suppl. 1):S26

Brandelli A (ed) (2021) Probiotics: advanced food and health applications. Academic Press

Broughton GI, Janis JE, Attinger CE (2006) Wound healing: an overview. Plast Reconstr Surg 117:1e-S–32e-S

CDC (2019) Global viral hepatitis: millions of people are affected. Ctr Dis Control Prev. https://www.cdc.gov/hepatitis/global/index.htm%0Ahttps://www.cdc.gov/hepatitis/global/index.htm%0Ahttps://www.cdc.gov/hepatitis/global/index.htm%0Ahttps://www.cdc.gov/hepatitis/global/index.htm#ref01

Chadwick VS, Chen W, Shu D, Paulus B, Bethwaite P, Tie A, Wilson I (2002) Activation of the mucosal immune system in irritable bowel syndrome. Gastroenterology 122:1778–1783

Chan JY, Yuen AC, Chan RY, Chan SW (2013) A review of the cardiovascular bene fits and antioxidant properties of allicin. Phytother Res 27:637–646

Chandrasekara A, Shahidi F (2018) Herbal beverages: bioactive compounds and their role in disease risk reduction: a review. J Tradit Complement Med 8(4):451–458

Chaughule RS, Barve RS (2023) Role of herbal medicines in the treatment of infectious diseases. Vegetos:1–11

Chen L, DiPietro LA (2017) Toll-like receptor function in acute wounds. Adv Wound Care 6:344–355

Chen P, Hong F, Yu X (2022) Prevalence of periodontal disease in pregnancy: a systematic review and meta-analysis. J Dent 125:104253. https://doi.org/10.1016/j.jdent.2022.104253

Chepulis L (2008) Healing honey: a natural remedy for better health and wellness. Universal-Publishers

Chew CY, Chua LS, Soontorngun N, Lee CT (2018) Discovering potential bioactive compounds from Tualang honey. Agric Nat Resources 52(4):361–365

Chithra P, Sajithlal GB, Chandrakasan G (1998) Influence of Aloe vera on collagen turnover in healing of dermal wounds in rats. Indian J Exp Biol 36(9):896–901

Choopani R, Sadr S, Kaveh S, Bayat H, Mosaddegh M (2017) Efficacy and safety of Iranian polyherbal formulation (compound honey syrup) in pediatric patients with mild to moderate asthma: a randomized clinical trial. Galen Med J 6(4):e884–e884

Church J (1954) Honey as a source of the anti-stiffness factor [abstract]. Fed Proc Am Physiol Soc 13(1):26

Cianciosi D, Forbes-Hernández TY, Afrin S, Gasparrini M, Reboredo-Rodriguez P, Manna PP et al (2018) Phenolic compounds in honey and their associated health benefits: a review. Molecules 23(9):2322

Clearwater MJ, Revell M, Noe S, Manley-Harris M (2018) Influence of genotype, floral stage, and water stress on floral nectar yield and composition of mānuka (*Leptospermum scoparium*). Ann Bot 121(3):501–512

Cohen D, Siegel A (2021) Ashkenazi herbalism: rediscovering the herbal traditions of Eastern European Jews. North Atlantic Books

Cohen SM, Purtilo DT, Ellwein LB (1991) Ideas in pathology. Pivotal role of increased cell proliferation in human carcinogenesis. Mod Pathol 4(3):371–382

Conti N, Cevenini G, Vannuccini S, Orlandini C, Valensise H, Gervasi MT, Ghezzi F, Di Tommaso M, Severi FM, Petraglia F (2014) Women with endometriosis at first pregnancy have an increased risk of adverse obstetric outcome. J Maternal-Fetal Neonatal Med 9:1–4

Cooper NP, Okolo S (2005) Fibroids in pregnancy—common but poorly understood. Obstet Gynecol Surv 60:132–138

Coppola N, Cantile T, Adamo D, Canfora F, Baldares S, Riccitiello F et al (2023) Supportive care and antiviral treatments in primary herpetic gingivostomatitis: a systematic review. Clin Oral Investig 27(11):6333–6344. https://doi.org/10.1007/s00784-023-05250-5

Couquet Y, Desmoulière A, Rigal ML (2013) The antibacterial and healing properties of honey. Pharm News 52(531):22–25

Cragg GM, Newman DJ (2013) Natural products: a continuing source of novel drug leads. Biochim Biophys Acta 1830(6):3670–3695

Dahiya D, Nigam PS (2022) The gut microbiota influenced by the intake of probiotics and functional foods with prebiotics can sustain wellness and alleviate certain ailments like gut-inflammation and colon-cancer. Microorganisms 10(3):665

Dassanayaka S, Jones SP (2015) Recent developments in heart failure. Circ Res 117(7):e58–e63. https://doi.org/10.1161/CIRCRESAHA.115.305765

Davison EK, Brimble MA (2019) A natural product derived privileged scaffolds in drug discovery. Curr Opin Chem Biol 52:1–8

De Padua LS, Bunyapraphatsara N, Lemmens RHMJ (1999) Plant resources of South-East Asia 12 (1) Medicinal and Poisonous plants 1. Backhuys Publishers, Leiden, The Netherlands

Deb S, Dasgupta A (2008) A study on risk factors of cardiovascular diseases in an urban health center of Kolkata. Indian J Commun Med 33(4):271–275. https://doi.org/10.4103/0970-0218.43239

Delavary BM, van der Veer WM, van Egmond M, Niessen FB, Beelen RH (2011) Macrophages in skin injury and repair. Immunobiology 216:753–762

Demeke CA, Woldeyohanins AE, Kifle ZD (2021) Herbal medicine use for the management of COVID-19: a review article. Metab Open 12:100141

Dhama K, Tiwari R, Chakraborty S, Saminathan M, Kumar A, Karthik K, Wani MY, Amarpal A, Singh SV, Rahal A (2014) Evidence based antibacterial potentials of medicinal plants and herbs countering bacterial pathogens especially in the era of emerging drug resistance: an integrated update. Int J Pharmacol 10(1):1–43

Din SRU, Saeed S, Khan SU, Kiani FA, Alsuhaibani AM, Zhong M (2023) Bioactive compounds (BACs): a novel approach to treat and prevent cardiovascular diseases. Curr Probl Cardiol 48(7):101664

Diretto G, Rubio-Moraga A, Argandoña J, Castillo P, Gómez-Gómez L, Ahrazem O (2017) Tissue-specific accumulation of sulfur compounds and saponins in different parts of garlic cloves from purple and white ecotypes. Molecules 22(8):1359

Dorgan JF, Baer DJ, Albert PS, Judd JT, Brown ED, Corle DK, Campbell WS, Hartman TJ, Tejpar AA, Clevidence BA, Giffen CA, Chandler DW, Stanczyk FZ, Taylor PR (2001) Serum hormones and the alcohol-breast cancer association in postmenopausal women. J Natl Cancer Inst 93:710–715

Drossman DA, Dumitrascu DL (2006) Rome III: new standard for functional gastrointestinal disorders. J Gastrointestin Liver Dis 15(3):237–241

Drossman DA, Creed FH, Olden KW, Svedlund J, Toner BB, Whitehead WE (1999) Psychosocial aspects of the functional gastrointestinal disorders. Gut 45(Suppl II):II25–II30. https://doi.org/10.1136/gut.45.2008.ii25

Dutta KN, Chetia P, Lahkar S, Das S (2014) Herbal plants used as diuretics: a comprehensive review. J Pharm Chem Biol Sci 2(1):27–32

Dżugan M, Sowa P, Kwaśniewska M, Wesołowska M, Czernicka M (2017) Physicochemical parameters and antioxidant activity of bee honey enriched with herbs. Plant Foods Hum Nutr 72:74–81

Ekor M (2014) The growing use of herbal medicines: issues relating to adverse reactions and challenges in monitoring safety. Front Pharmacol 4:177

Ellulu MS, Patimah I, Khaza'ai H, Rahmat A, Abed Y (2017) Obesity and inflammation: the linking mechanism and the complications. Arch Med Sci 13(4):851–863

El-Saber Batiha G, Magdy Beshbishy A, Wasef LG, Elewa YHA, Al-Sagan AA, Abd El-Hack ME, Prasad Devkota H (2020) Chemical constituents and pharmacological activities of garlic (Allium sativum L.): a review. Nutrients 12(3):872

El-Seedi HR, Eid N, Abd El-Wahed AA, Rateb ME, Afifi HS, Algethami AF et al (2022) Honey bee products: preclinical and clinical studies of their anti-inflammatory and immunomodulatory properties. Front Nutr 8:761267

Erejuwa OO, Sulaiman SA, Wahab MSA, Sirajudeen KNS, Salleh MSM, Gurtu S (2011) Differential responses to blood pressure and oxidative stress in streptozotocin-induced diabetic wistar-kyoto rats and spontaneously hypertensive rats: effects of antioxidant (honey) treatment. Int J Mol Sci 12(12):1888–1907. https://doi.org/10.3390/ijms12031888

Erejuwa OO, Sulaiman SA, Ab Wahab MS (2012a) Honey-a novel antidiabetic agent. Int J Biol Sci 8(6):913–934. https://doi.org/10.7150/ijbs.3697

Erejuwa OO, Sulaiman SA, Ab Wahab MS (2012b) Honey: a novel antioxidant. Molecules 17(4):4400–4423

Erejuwa OO, Sulaiman SA, Wahab MSA (2014) Effects of honey and its mechanisms of action on the development and progression of cancer. Molecules 19:2497–2522

Eteraf-Oskouei T, Najafi M (2013) Traditional and modern uses of natural honey in human diseases: a review. Iran J Basic Med Sci 16(6):731. PMCID: PMC3758027

Ewnetu Y, Lemma W, Birhane N (2014) Synergetic antimicrobial effects of mixtures of Ethiopian honeys and ginger powder extracts on standard and resistance clinical bacteria isolates. Evid Based Complement Alternat Med 2014:562804. https://doi.org/10.1155/2014/562804

Fabricant DS, Farnsworth NR (2001) The value of plants used in traditional medicine for drug discovery. Environ Health Perspect 109(Suppl. 1):69–75

Fadzil MAM, Mustar S, Rashed AA (2023) The potential use of honey as a neuroprotective agent for the management of neurodegenerative diseases. Nutrients 15(7):1558. https://doi.org/10.3390/nu15071558

Farooqui T, Farooqui A, A. (2011) Health benefits of honey: implications for treating cardiovascular diseases. Curr Nutr Food Sci 7(4):232–252

Fauzi AN, Norazmi MN, Yaacob NS (2011) Tualang honey induces apoptosis and disrupts the mitochondrial membrane potential of human breast and cervical cancer cell lines. Food Chem Toxicol 49(4):871–878

Feknous N, Boumendjel M (2022) Natural bioactive compounds of honey and their antimicrobial activity. Czech J Food Sci 40(3):163–178

Fell JW (1857) Treatise on cancer and its treatment. J. Churchill, London

Ferlay J, Soerjomataram I, Dikshit R, Eser S, Mathers C, Rebelo M, Parkin DM, Forman D, Bray F (2014) Cancer incidence and mortality worldwide: sources, methods and major patterns in GLOBOCAN. Int J Cancer 136:E359–E386

Forleo C, Carmosino M, Resta N, Rampazzo A, Valecce R, Sorrentino S, Iacoviello M, Pisani F, Procino G, Gerbino A et al (2015) Clinical and functional characterization of a novel mutation in lamin a/c gene in a multigenerational family with arrhythmogenic cardiac laminopathy. PLoS One 10:e0121723

Foroozandeh M, Hosseini SE (2017) Effects of metformin and ginger rhizome extract on the pituitary - gonad function in adult female rats with polycystic ovary syndrome. Armaghane Danesh J 22(3):337–349

Fossum G (2014) Assessment report on *Sambucus nigra* L., fructus. European Medicines Agency, London, pp 1–25

Foster GR, Goldin RD, Thomas HC, Owens DK (1998) Chronic hepatitis C virus infection causes a significant reduction in quality of life in the absence of cirrhosis. Hepatology 27(1):209–212. https://doi.org/10.1002/hep.510270132

Frishman WH, Beravol P, Carosella C (2009) Alternative and complementary medicine for preventing and treating cardiovascular disease. Dis Mon 55(3):121–192

Fuchs FD, Whelton PK (2019) High blood pressure and cardiovascular disease. Hypertension 75(2):285–292. https://doi.org/10.1161/HYPERTENSIONAHA.119.14240

Galiano RD, Mustoe TA (2007) Wound care. In: Thorne CH, Beasley RW, Aston SJ et al (eds) Grabb and Smith's plastic surgery, 6th edn. Lippincott Williams & Wilkins, Philadelphia, PA, pp 23–32

Gautam MK et al (2014) In vivo healing potential of Aegle marmelos in excision, incision, and dead space wound models. Sci World J 2014:740107

Genco RJ, Borgnakke WS (2013) Risk factors for periodontal disease. Periodontol 2000 62(1):59–94. https://doi.org/10.1111/j.1600-0757.2012.00457.x

Ghagane SC, Akbar AA (2023) Use of honey in cardiovascular diseases. In: Honey: composition and health benefits. Wiley, pp 197–209

Ghasemian M, Owlia S, Owlia MB (2016) Review of anti-inflammatory herbal medicines. Adv Pharmacol Pharm Sci 2016:9130979. https://doi.org/10.1155/2016/9130979

Ghashm AA, Othman NH, Khattak MN, Ismail NM, Saini R (2010) Antiproliferative effect of Tualang honey on oral squamous cell carcinoma and osteosarcoma cell lines. BMC Complement Alternat Med 10:1–8

Goel A, Kunnumakkara AB, Aggarwal BB (2008) Curcumin as "Curecumin": from kitchen to clinic. BiochemPharmacol 75:787–809

Golebiewska EM, Poole AW (2015) Platelet secretion: from haemostasis to wound healing and beyond. Blood Rev 29:153–162

Goodman SG, Steg PG, Eagle KA, Fox KA, Lopez-Sendon J, Montalescot G, Budaj A, Kennelly BM, Gore JM, Allegrone J, Granger CB, Gurfinkel EP, Investigators GRACE (2006) The diagnostic and prognostic impact of the redefinition of acute myocardial infarction: lessons from the Global Registry of Acute Coronary Events (GRACE). Am Heart J 151(3):654–660

Gorjanović SŽ, Alvarez-Suarez JM, Novaković MM, Pastor FT, Pezo L, Battino M, Sužnjević DŽ (2013) Comparative analysis of antioxidant activity of honey of different floral sources using recently developed polarographic and various spectrophotometric assays. J Food Compos Anal 30(1):13–18

Gośliński M, Nowak D, Szwengiel A (2021) Multidimensional comparative analysis of bioactive phenolic compounds of honey of various origins. Antioxidants 10(4):530. https://doi.org/10.3390/antiox10040530

Grabek-Lejko D, Miłek M, Sidor E, Puchalski C, Dżugan M (2022) Antiviral and antibacterial effect of honey enriched with *Rubus* spp. as a functional food with enhanced antioxidant properties. Molecules 27(15):4859. https://doi.org/10.3390/molecules27154859

Greene MW, Burrington CM, Lynch DT, Davenport SK, Johnson AK, Horsman MJ, Chowdhry S, Zhang J, Sparks JD, Tirrell PC (2014) Lipid metabolism, oxidative stress and cell death are regulated by PKC delta in a dietary model of nonalcoholic steatohepatitis. PLoS One 9:e85848

Greenwell M, Rahman PKSM (2015) Medicinal plants: their use in anticancer treatment. Int J Pharm Sci Res 6(10):4103–4112. https://doi.org/10.13040/IJPSR.0975-8232.6(10).4103-12

Grigoleit HG, Grigoleit P (2005) Gastrointestinal clinical pharmacology of peppermint oil. Phytomedicine 12:607–611

Gruhlke MC, Nwachwukwu I, Arbach M, Anwar A, Noll U, Slusarenko AJ (2011) Allicin from garlic, effective in controlling several plant diseases, is a reactive sulfur species (RSS) that pushes cells into apoptosis. In: Modern fungicides and antifungal compounds VI. 16th International Reinhardsbrunn Symposium, Friedrichroda, Germany, April 25–29, 2010. Deutsche Phytomedizinische Gesellschaft eV Selbstverlag, pp 325–330

Guimarães R, Barros L, Carvalho AM, Ferreira IC (2011) Infusions and decoctions of mixed herbs used in folk medicine: synergism in antioxidant potential. Phytother Res 25(8):1209–1214

Gunson RN, Shouval D, Roggendorf M, Zaaijer H, Nicholas H, Holzmann H, De Schryver A, Reynders D, Connell J, Gerlich WH, Marinho RT, Tsantoulas D, Rigopoulou E, Rosenheim M, Valla D, Puro V, Struwe J, Tedder R, Aitken C et al (2003) Hepatitis B virus (HBV) and hepatitis C virus (HCV) infections in health care workers (HCWs): guidelines for prevention of transmission of HBV and HCV from HCW to patients. J Clin Virol 27(3):213–230. https://doi.org/10.1016/S1386-6532(03)00087-8

Gwee KA, Collins SM, Read NW, Rajnakova A, Deng Y, Graham JC, McKendrick MW, Moochhala SM (2003) Increased rectal mucosal expression of interleukin 1beta in recently acquired post-infectious irritable bowel syndrome. Gut 52:523–526

Hadi IH (2017) Effect of honey on sperm characteristics and pregnancy rate in mice. Bullet Iraq Nat History Museum 14:223–233

Han W, Wang ZJ, Zhao B, Yang XQ, Wang D, Wang JP et al (2005) Pathologic change of elastic fibers with difference of microvessel density and expression of angiogenesis-related proteins in internal hemorrhoid tissues. J Gastrointest Surg 8(1):56–59. https://doi.org/10.1111/j.1445-2197.2005.03280.x

Hashim KN, Chin KY, Ahmad F (2021) The mechanism of honey in reversing metabolic syndrome. Molecules 26(4):808. https://doi.org/10.3390/molecules26040808

Hassan MI, Mabrouk GM, Shehata HH, Aboelhussein MM (2012) Antineoplastic effects of bee honey and *Nigella sativa* on hepatocellular carcinoma cells. Integr Cancer Ther 11(4):354–363

Hegazi A, Abdou M, Allah FA (2013) Influence of honey on immune response against Newcastle disease vaccine. Int J Basic Appl Virol 2:1–5

Hemayatkhah-Jahromi V, Rahmanian Koushkaki M (2016) Effect of hydro-alcoholic extract of Aloe vera L. on polycystic ovary syndrome in rat. Feyz Med Sci J 20(3):221–227

Henderson BE, Ross RK, Pike MC, Casagrande JT (1982) Endogenous hormones as a major factor in human cancer. Cancer Res 42(8):3232–3239

Henriques AF, Jenkins RE, Burton NF et al (2010) The intracellular effects of manuka honey on Staphylococcus aureus. Eur J Clin Microbiol Infect Dis 29(1):45–50

Hever J (2016) Plant-based diets: a physician's guide. Perm J 20(3):93–101

Hewlings SJ, Kalman DS (2017) Curcumin: a review of its effects on human health. Foods 6(10):92

Hidayati T, Indrayanti I, Darmawan E, Akrom A (2023) Herbal honey preparations of *Curcuma xanthorriza* and black cumin protect against carcinogenesis through antioxidant and immuno-modulatory activities in Sprague Dawley (SD) rats induced with dimethylbenz (a) anthracene. Nutrients 15(2):371

Hofmeister MG, Foster MA, Teshale EH (2019) Epidemiology and transmission of hepatitis a virus and hepatitis e virus infections in the United States. Cold Spring Harb Perspect Med 9(4):a033431. https://doi.org/10.1101/cshperspect.a033431

Hosseini SE, Mehrabani D, Razavi F (2014) Effect of palm pollen extract on sexual hormone levels and follicle numbers in adult female BALB/c mice. Horizon Med Sci J 20(3):139–143

Hostettmann K (2003) History of a plant: the example of Echinacea. Res Complement Nat Classical Med 10:9–12

Ibrahim MA, Berahim Z, Ahmad A, Taib H (2021) The effect of locally delivered Tualang honey on healing of periodontal tissues during non-surgical periodontal therapy. IIUM J Orofacial Health Sci 2(2):16–26. https://doi.org/10.31436/ijohs.v2i2.65

Inamdar AA, Inamdar AC (2016) Heart Failure: diagnosis, management and utilization. J Clin Med 5(7):62. https://doi.org/10.3390/jcm5070062

Isidorov VA, Bagan R, Bakier S, Swiecicka I (2015) Chemical composition and antimicrobial activity of Polish herb honeys. Food Chem 171:84–88

Islam MA (2012) Cardiovascular effects of green tea catechins: progress and promise. Recent Pat Cardiovasc Drug Discov 7(2):88–99

Israili ZH (2014) Antimicrobial properties of honey. Am J Ther 21(4):304–323

Jafari M, Goldasteh S, Aghdam HR, Zamani AA, Soleyman-Nejadian E, Schausberger P (2023) Modeling thermal developmental trajectories and thermal requirements of the ladybird *Stethorus gilvifrons*. Insects 14(7):581. https://doi.org/10.3390/insects14070581

Jaganathan SK, Mandal M (2010) Involvement of non-protein thiols, mitochondrial dysfunction, reactive oxygen species and p53 in honey-induced apoptosis. Invest New Drugs 28:624–633. https://doi.org/10.1007/s10637-009-9302-0

Jalili C, Taghadosi M, Pazhouhi M, Bahrehmand F, Miraghaee SS, Pourmand D, Rashidi I (2020) An overview of therapeutic potentials of Taraxacum officinale (dandelion): a traditionally valuable herb with a reach historical background. World Cancer Res J 7:e1679

James C, Harfouche M, Welton NJ, Turner ME, Abu-raddad LJ (2020) Herpes simplex virus: global infection prevalence and incidence estimates, 2016. Bull World Health Organ 98(5):315–329. https://doi.org/10.2471/BLT.19.237149

Javadi SMR, Hashemi M, Mohammadi Y, MamMohammadi A, Sharifi A, Makarchian HR (2018) Synergistic effect of honey and Nigella sativa on wound healing in rats. Acta Cir Bras 33:518–523

Johari H (1994) The healing cuisine. Healing Arts Press, Rochester, VT

Johnson BK, Abramovitch RB (2017) Small molecules that sabotage bacterial virulence. Trends Pharmacol Sci 38:339–362

Johnston M, McBride M, Dahiya D, Owusu-Apenten R, Nigam PS (2018) Antibacterial activity of Manuka honey and its components: an overview. AIMS Microbiol 4(4):655–664

Jordan MA (2002) Mechanism of action of antitumor drugs that interact with microtubules and tubulin. Curr Med Chem-Anti-Cancer Agents 2(1):1–17

Josso N, Picard JY, Imbeaud S, Carre ED, Zeller J, Adamsbaum C (1993) The persistent müllerian duct syndrome: a rare cause of cryptorchidism. Eur J Pediatr 152(Suppl. 2):S76–S78

Kaiser P, Wächter J, Windbergs M (2021) Therapy of infected wounds: overcoming clinical challenges by advanced drug delivery systems. Drug Deliv Transl Res 11:1545–1567

Kalan L, Loesche M, Hodkinson BP, Heilmann K, Ruthel G, Gardner SE, Grice EA (2016) Redefining the chronic-wound microbiome: fungal communities are prevalent, dynamic, and associated with delayed healing. MBio 7:e01058–e01016

Kalfa N, Philibert P, Sultan C (2009) Is hypospadias a genetic, endocrine or environmental disease, or still an unexplained malformation? Int J Androl 32:187–197

Kannel WB (1976) Some lessons in cardiovascular epidemiology from Framingham. Am J Cardiol 37:269–282. https://doi.org/10.1016/0002-9149(76)90323-4

Khalil ML, Sulaiman SA (2010) The potential role of honey and its polyphenols in preventing heart disease: a review. Afr J Tradit Complement Alternat Med 7(4):315–321

Khalil MI, Tanvir E, Afroz R, Sulaiman SA, Gan SH (2015) Cardioprotective effects of Tualang honey: amelioration of cholesterol and cardiac enzymes levels. Biomed Res Int 286051. https://doi.org/10.1155/2015/286051

Khan MJ, Ullah A, Basit S (2019) Genetic basis of polycystic ovary syndrome (PCOS): current perspectives. Appl Clin Genet 12:249–260

Kim YS, Kim JW, Ha NY, Kim J, Ryu HS (2020) Herbal therapies in functional gastrointestinal disorders: a narrative review and clinical implication. Front Psych 11:601. https://doi.org/10.3389/fpsyt.2020.00601

Kingma J, Simard C, Drolet B (2023) Overview of cardiac arrhythmias and treatment strategies. Pharmaceuticals 16:844. https://doi.org/10.3390/ph16060844

Kingsley K, Huff J, Rust W, Carroll K, Martinez A, Fitchmun M, Plopper GE (2002) ERK1/2 mediates PDGF-BB stimulated vascular smooth muscle cell proliferation and migration on laminin-5. Biochem Biophys Res Commun 293:1000–1006

Kolaczkowska E, Kubes P (2013) Neutrophil recruitment and function in health and inflammation. Nat Rev Immunol 13:159–175

Kotb MA, Kotb MA, Abdalla AK (2014) Sustained Hepatitis C virus clearance was achieved by honey based conservative management in 35% of chronic Hepatitis C virus patients: a prospective. Interferon 82(1):639–645. https://www.researchgate.net/profile/Magd-Kotb/publication/275658952_Sustained_Hepatitis_C_Virus_Clearance_was_Achieved_by_Honey_Based_Conservative_Management_in_35_of_Chronic_Hepatitis_C_Virus_Patients_A_Prospective_Cohort_Study/links/5599bcd108ae21086d25b550/Sustained-Hepatitis-C-Virus-Clearance-was-Achieved-by-Honey-Based-Conservative-Management-in-35-of-Chronic-Hepatitis-C-Virus-Patients-A-Prospective-Cohort-Study.pdf

Krishnaiah D, Sarbatly R, Bono A (2007) Phytochemical antioxidants for health and medicine—a move towards nature. Biotechnol Mol Biol Rev 1:97–104

Kujath P, Michelsen A (2008) Wounds-from physiology to wound dressing. Deutsches Arzteblatt Int 105(13):239–248

Kumar KS, Bhowmik D, Biswajit C, Chandira MR (2010a) Medicinal uses and health benefits of honey: an overview. J Chem Pharm Res 2(1):385–395

Kumar S, Malhotra R, Kumar D (2010b) *Euphorbia hirta*: its chemistry, traditional and medicinal uses, and pharmacological activities. Pharmacogn Rev 4(7):58

Kumar S, Verma M, Hajam YA, Kumar R (2024) Honey infused with herbs: a boon to cure pathological diseases. Heliyon 10(1):E23302. https://doi.org/10.1016/j.heliyon.2023.e23302

Kumari I, Sharma A, Sharma P, Hajam YA, Kumar R (2021) Composition of honey, its therapeutic properties and role in cosmetic industry. In: Honey. CRC Press, pp 157–178

Kurek-Górecka A, Górecki M, Rzepecka-Stojko A, Balwierz R, Stojko J (2020) Bee products in dermatology and skin care. Molecules 25(3):556. https://doi.org/10.3390/molecules25030556

Lai J, Caughey AB, Qidwai GI, Jacoby AF (2012) Neonatal outcomes in women with sonographically identified uterine leiomyomata. J Matern-Fetal Neonatal Med 25:710–713

Laksemi DAAS, Tunas IK, Damayanti PAA, Sudarmaja IM, Widyadharma IPE, Wiryanthini IAD, Linawati NM (2023) Evaluation of antimalarial activity of combination extract of Citrus aurantifolia and honey against plasmodium berghei-İnfected mice. Trop J Nat Prod Res 7(1):2168–2171. https://doi.org/10.26538/tjnpr/v7i1.13

Leitch A (1923) A British medical association lecture on the experimental inquiry into the causes of cancer. Br Med J 2:1–7

Lewington S, Clarke R, Qizilbash N, Peto R, Collins R, Prospective Studies Collaboration (2002) Age-specific relevance of usual BP to vascular mortality: a meta-analysis of individual data for one million adults in 61 prospective studies. Lancet 360:1903–1913. https://doi.org/10.1016/s0140-6736(02)11911-8

Liu C-F, Lin C-C, Lin M-H, Lin Y-S, Lin S-C (2002) Cytoprotection by propolis ethanol extract of acute absolute ethanol-induced gastric mucosal lesions. Am J Chin Med 30(2-3):245–254

Lobo V, Patil A, Phatak A, Chandra N (2010) Free radicals, antioxidants and functional foods: impact on human health. Pharmacogn Rev 4(8):118

Lopez-Lazaro MA (2016) Local mechanism by which alcohol consumption causes cancer. Oral Oncol 62:149–152. https://doi.org/10.1016/j.oraloncology.2016.10.001

Lu WI, Lu DP (2014) Impact of Chinese herbal medicine on American society and health care system: perspective and concern. Evid Based Complement Alternat Med 2014:251891

Luchese RH, Prudêncio ER, Guerra AF (2017) Honey as a functional food. Honey Anal:287–307

Lukasiewicz M, Kowalski S, Makarewicz M (2015) Antimicrobial and antioxidant activity of selected Polish herb honeys. LWT-Food Sc Technol 64(2):547–553

Lukic J, Chen V, Strahinic I, Begovic J, Lev-Tov H, Davis SC, Tomic-Canic M, Pastar I (2017) Probiotics or pro-healers: the role of beneficial bacteria in tissue repair. Wound Repair Regen 25:912–922

Lusby PE, Coombes A, Wilkinson JM (2002) Honey: a potent agent for wound healing? J WOCN 29(6):295–300

Lusby PE, Coombes AL, Wilkinson JM (2005) Bactericidal activity of different honeys against pathogenic bacteria. Arch Med Res 36(5):464–467

Mackin C, Dahiya D, Nigam PS (2023) Honey as a natural nutraceutical: its combinational therapeutic strategies applicable to blood infections—Septicemia, HIV, SARS-CoV-2, Malaria. Pharmaceuticals 16(8):1154. https://doi.org/10.3390/ph16081154

Maddocks SE, Jenkins RE (2013) Honey: a sweet solution to the growing problem of antimicrobial resistance? Future Microbiol 8(11):1419–1429

Madhivanan P, Krupp K, Hardin J, Karat C, Klausner JD, Reingold AL (2009) Simple and inexpensive point-of-care tests improve diagnosis of vaginal infections in resource constrained settings. Trop Med Int Health 14(6):703–708. https://doi.org/10.1111/j.1365-3156.2009.02274.x. Epub 2009 Apr 20. PMID: 19392745; PMCID: PMC3625926

Majerus M-A (2022) The cause of cancer: the unifying theory. Adv Cancer Biol Metastasis 4:100034

Malek SNA, Phang CW, Ibrahim H, Wahab NA, Sim KS (2011) Phytochemical and cytotoxic investigations of *Alpinia mutica* rhizomes. Molecules 16(1):583–589

Malfertheiner P, Chan FK, McColl KE (2009) Peptic ulcer disease. Lancet 374(9699):1449–1461

Mamun MA, Absar N (2018) Major nutritional compositions of black cumin seeds–cultivated in Bangladesh and the physicochemical characteristics of its oil. Int Food Res J 25(6):2634–2639

Mandal MD, Mandal S (2011) Honey: its medicinal property and antibacterial activity. Asian Pac J Trop Biomed 1(2):154–160. https://doi.org/10.1016/S2221-1691(11)60016-6

Mao W, Liao X, Wu W, Yu Y, Yang G (2017) The clinical characteristics of patients with chronic idiopathic anal pain. Open Med 12(1):92–98. https://doi.org/10.1515/med-2017-0015

Marenberg ME, Risch N, Berkman LF, Floderus B, de Faire U (1994) Genetic susceptibility to death from coronary heart disease in a study of twins. N Engl J Med 330:1041–1046

Martin KW, Ernst E (2003) Antiviral agents from plants and herbs: a systematic review. Antivir Ther 8(2):77–90

Martinez-Armenta C, Camacho-Rea MC, Martínez-Nava GA, Espinosa-Velázquez R, Pineda C, Gomez-Quiroz LE, López-Reyes A (2021) Therapeutic potential of bioactive compounds in honey for treating osteoarthritis. Front Pharmacol 12:642836

Masad RJ, Haneefa SM, Mohamed YA, Al-Sbiei A, Bashir G, Fernandez-Cabezudo MJ, Al-Ramadi BK (2021) The immunomodulatory effects of honey and associated flavonoids in cancer. Nutrients 13(4):1269

Massart F, Saggese G (2010) Morphogenetic targets and genetics of undescended testis. Sex Dev 4:326–335

Massoura, M. (2020). An investigation into the antibacterial mechanism of honey. (Doctoral dissertation, University of Birmingham)

Mazumder A, Kumar N, Das S (2023) A comparative evaluation of various therapies of synthetic drugs with amla and honey combination for the treatment of gastroesophageal reflux disease. Ind J Pharm Edu Res 57(2):540–546

Mearin F, Perez-Oliveras M, Perello A, Vinyet J, Ibanez A, Coderch J, Perona M (2005) Dyspepsia after a Salmonella gastroenteritis outbreak: one-year follow-up cohort study. Gastroenterology 129:98–104

Miłek M, Ciszkowicz E, Sidor E, Hęclik J, Lecka-Szlachta K, Dżugan M (2023) The antioxidant, antibacterial and anti-biofilm properties of rapeseed creamed honey enriched with selected plant superfoods. Antibiotics 12(2):235. https://doi.org/10.3390/antibiotics12020235

Miller AL (1998) Botanical influences on cardiovascular disease. Altern Med Rev 706(6):422–431

Miner J, Hoffhines A (2007) The discovery of aspirin's antithrombotic effects. Tex Heart Inst J 34(2):179

Moalli PA, MacDonald JL, Goodglick LA, Kane AB (1987) Acute injury and regeneration of the mesothelium in response to asbestos fibers. Am J Pathol 128(3):426–445

Mobeen S, Ravindra SV, Sunitha JD, Prakash R, Satyanarayana D, Swayampakula H et al (2023) A novel herbal paste formulation of turmeric, tulsi, and honey for the treatment of oral submucous fibrosis. Cureus 15(10):e46608. https://doi.org/10.7759/cureus.46608

Modaresi M, Messripour M, Rajaei R (2009) The effect of cinnamon (Bark) extract on male reproductive physiology in mice. Armaghane Danesh J 14(1):67–77

Mohamed M, Sulaiman SA, Jaafar H, Sirajudeen KN (2011) Antioxidant protective effect of honey in cigarette smoke-induced testicular damage in rats. Int J Mol Sci 12:5508–5521

Mohamed M, Sulaiman SA, Sirajudeen KN (2013) Protective effect of honey against cigarette smoke induced-impaired sexual behavior and fertility of male rats. Toxicol Ind Health 29:264–271

Moniruzzaman M, Khalil M, Sulaiman S, Gan S (2012) Advances in the analytical methods for determining the antioxidant properties of honey: a review. Afr J Tradit Complement Alternat Med 9:36e42

Mosavat M, Ooi FK, Mohamed M (2014) Stress hormone and reproductive system in response to honey supplementation combined with different jumping exercise intensities in female rats. Biomed Res Int 2014:1–6

Mračević SĐ, Krstić M, Lolić A, Ražić S (2020) Comparative study of the chemical composition and biological potential of honey from different regions of Serbia. Microchem J 152:104420

Mullaicharam AR, Halligudi N (2019) St John's wort (Hypericum perforatum L.): a review of its chemistry, pharmacology and clinical properties. Int J Res Phytochem Pharmacol Sci 1(1):5–11

Mumtaz PT, Bashir SM, Rather MA, Dar KB, Taban Q, Sajood S et al (2020) Antiproliferative and apoptotic activities of natural honey. Ther Appl Honey Phytochem 1:345–360

Munakata J, Naliboff B, Harraf F, Kodner A, Lembo T, Chang L, Silverman DH, Mayer EA (1997) Repetitive sigmoid stimulation induces rectal hyperalgesia in patients with irritable bowel syndrome. Gastroenterology 112:55–63

Münstedt K (2019) Bee products and the treatment of blister-like lesions around the mouth, skin and genitalia caused by herpes viruses—a systematic review. Complement Ther Med 43:81–84. https://doi.org/10.1016/j.ctim.2019.01.014

Mustar S, Ibrahim N (2022) A sweeter pill to swallow: a review of honey bees and honey as a source of probiotic and prebiotic products. Foods 11(14):2102

Nabavi SF, Maggi F, Daglia M, Habtemariam S, Rastrelli L, Nabavi SM (2016) Pharmacological effects of *Capparis spinosa* L. Phytother Res 30(11):1733–1744

Nadkarni AK (2007) Dr. KM Nadkarni's Indian materia medica: with Ayurvedic, Unani-tibbi, Siddha, allopathic, homeopathic, naturopathic & home remedies, appendices & indexes, vol 1. Popular Prakashan

Nag SA, Qin JJ, Wang W, Wang MH, Zhang R (2012) Ginsenosides as anticancer agents: in vitro and in vivo activities, structure-activity relationships, and molecular mechanisms of action. Front Pharmacol 3(25):1–18

Naik S et al (2015) Commensal–dendritic-cell interaction specifies a unique protective skin immune signature. Nature 520:104–108

Naik PP, Mossialos D, van Wijk B, Novakova P, Wagener FADTG, Cremers NAJ (2021) Medical-grade honey outperforms conventional treatments for healing cold sores—a clinical study. Pharmaceuticals 14(12):1264. https://doi.org/10.3390/ph14121264

Nassar D, Blanpain C (2016) Cancer stem cells: basic concepts and therapeutic implications. Annu Rev Pathol 11:47–76

Nazir S, Rafique N, Ahad K (2017) Comparative Evaluation of Extraction Procedures and Chromatographic Techniques for Analysis of Multiresidue Pesticides in Honey. J Environ Toxicol Stu 1(1)

Nazir M, Al-Ansari A, Al-Khalifa K, Alhareky M, Gaffar B, Almas K (2020) Global prevalence of periodontal disease and lack of its surveillance. Sci World J 2020. https://doi.org/10.1155/2020/2146160

Ness J, Sherman FT, Pan CX (1999) Alternative medicine: what the data say about common herbal therapies. Geriatrics (Basel, Switzerland) 54(10):33–38

Ng QX, Venkatanarayanan N, Ho CYX (2017) Clinical use of Hypericum perforatum (St John's wort) in depression: a meta-analysis. J Affect Disord 210:211–221

Nicoll R, Henein MY (2009) Ginger (Zingiber officinale Roscoe): a hot remedy for cardiovascular disease. Int J Cardiol 131(3):408–409

Nikhat S, Fazil M (2022) History, phytochemistry, experimental pharmacology and clinical uses of honey: a comprehensive review with special reference to Unani medicine. J Ethnopharmacol 282:114614

Noor-E-Tabassum DR, Lami MS, Chakraborty AJ, Mitra S, Tallei TE, Idroes R, Mohamed AA, Hossain MJ, Dhama K, Mostafa-Hedeab G, Emran TB (2022) Ginkgo biloba: a treasure of functional phytochemicals with multimedicinal applications. Evid Based Complement Alternat Med 2022:8288818. https://doi.org/10.1155/2022/8288818. PMID: 35265150; PMCID: PMC8901348

Norman RJ, Noakes M, Wu R, Davies MJ, Moran L, Wang JX (2004) Improving reproductive performance in overweight/obese women with effective weight management. Hum Reprod Update 10:267–280

Nunes A, Sousa M (2011) Use of valerian in anxiety and sleep disorders: what is the best evidence. Acta Med Port 24:961–966

Nyrop KA, Palsson OS, Levy RL, Von Korff M, Feld AD, Turner MJ, Whitehead WE (2007) Costs of health care for irritable bowel syndrome, chronic constipation, functional diarrhoea and functional abdominal pain. Aliment Pharmacol Ther 26:237–248

Ody P (2017) The complete medicinal herbal: a practical guide to the healing properties of herbs. Simon and Schuster

Ohtani T, Mohammed SF, Yamamoto K (2012) Diastolic stiffness as assessed by diastolic wall strain is associated with adverse remodeling and poor outcomes in heart failure with preserved ejection fraction. Eur Heart J 33:1742–1749

Okhiria OA, Henriques AFM, Burton NF et al (2009) Honey modulates biofilms of Pseudomonas aeruginosa in a time and dose dependent manner. J ApiProduct ApiMed Sci 1(1):6–10

Olaitan PB, Adeleke OE, Iyabo OO (2007) Honey: a reservoir for microorganisms and an inhibitory agent for microbes. Afr Health Sci 7(3):159–165

Olokoba AB, Obateru OA, Olokoba LB (2012) Type 2 diabetes mellitus: a review of current trends. Oman Med J 27(4):269–273. https://doi.org/10.5001/omj.2012.68

Onyeaghala AA, Anyiam AF, Husaini DC, Onyeaghala EO, Obi E (2023) Herbal supplements as treatment options for COVID-19: a call for clinical development of herbal supplements for emerging and re-emerging viral threats in Sub-Saharian Africa. Sci Afr 20:e01627

Onyeka IP, Bako SP, Suleiman MM, Onyegbule FA, Morikwe UC, Ogbue CO (2020) Antiulcer effects of methanol extract of *Euphorbia hirta* and honey combination in rats. Biomed Res Int:6827504. https://doi.org/10.1155/2020/6827504

Ortmans S, Daval C, Aguilar M, Compagno P, Cadrin-Tourigny J, Dyrda K, Rivard L, Tadros R (2019) Pharmacotherapy in inherited and acquired ventricular arrhythmia in structurally normal adult hearts. Expert Opin Pharmacother 20(17):2101–2114

Pan SY, Zhou SF, Gao SH, Yu ZL, Zhang SF, Tang MK et al (2013) New perspectives on how to discover drugs from herbal medicines: CAM's outstanding contribution to modern therapeutics. Evid Based Complement Alternat Med 2013:627375. https://doi.org/10.1155/2013/627375

Parastaka G, Nihayati E (2019) Respon pertumbuhan planlet tanaman Temulawak (Curcuma xanthorrhiza Roxb.) klon Jember dan Pasuruan terhadap berbagai konsentrasi kolkisin. J Produksi Tanaman 7(3):261–267

Park JS, Kim J, Elghiaty A, Ham WS (2018) Recent global trends in testicular cancer incidence and mortality. Medicine (Baltimore) 97(37):e12390

Pasupuleti VR, Arigela CS, Gan SH, Salam SKN, Krishnan KT, Rahman NA, Jeffree MS (2020) A review on oxidative stress, diabetic complications, and the roles of honey polyphenols. Oxid Med Cell Longev 2020:8878172. https://doi.org/10.1155/2020/8878172

Pattanayak P et al (2010) *Ocimum sanctum* Linn. A reservoir plant for therapeutic applications: an overview. Pharmacogn Rev 4(7):95–105. https://doi.org/10.4103/0973-7847.65323

Pavlova T, Stamatovska V, Kalevska T, Dimov I, Assistant G, Nakov G (2018) Quality characteristics of honey: a review, vol 57. Proc Univ Ruse, p 42

Peddireddy V, Siva Prasad B, Gundimeda S, Penagaluru P, Mundluru H (2012) Assessment of 8-oxo-7, 8-dihydro-2'-deoxyguanosine and malondialdehyde levels as oxidative stress markers and antioxidant status in non-small cell lung cancer. Biomarkers 17:261–268

Petrovska BB (2012) Historical review of medicinal plants' usage. Pharmacogn Rev 6(11):1

Pichichero E, Cicconi R, Mattei M, Muzi M, Canini A (2010) Acacia honey and chrysin reduce proliferation of melanoma cells through alterations in cell cycle progression. Int J Oncol 37:973–981

Plummer M, de Martel C, Vignat J, Ferlay J, Bray F, Franceschi S (2016) Global burden of cancers attributable to infections in 2012: a synthetic analysis. Lancet Glob Health 4(9):e609–e616

Poswal FS, Russell G, Mackonochie M, MacLennan E, Adukwu EC, Rolfe V (2019) Herbal teas and their health benefits: a scoping review. Plant Foods Hum Nutr 74:266–276

Poursaleh Z, Choopani R, Vahedi E, Khedmat AF, Ghazvini A, Salesi M, Ghanei M (2022) Effect of herbal medicine formulation (compound honey syrup) on quality of life in patients with COPD: a randomized clinical trial. Tanaffos 21(3):336

Prakash P, Gupta N (2005) Therapeutic uses of *Ocimum sanctum* Linn (Tulsi) with a note on eugenol and its pharmacological actions: a short review. Indian J Physiol Pharmacol 49(2):125

Pruthi JS (1976) Spices and condiments. National Book Trust, India

Purohit SK, Solanki R, Soni MK, Mathur V (2012) Experimental evaluation of Aloe vera leaves pulp as topical medicament on wound healing. Int J Pharmacol Res 2(3):110–112

Putri NM, Lumbuun RFM, Kreshanti P, Saharman YR, Tunjung N (2022) Comparison study of bacterial profile, wound healing, and cost effectiveness in pressure injury patients using treatment honey dressing and hydrogel. Jurnal Plastik Rekonstruksi 9(1):30–41. https://doi.org/10.14228/jprjournal.v9i1.334

Qadir J, Wani JA, Ali S, Yatoo AM, Zehra U, Rasool S et al (2020) Heath benefits of phenolic compounds in honey: an essay. Ther Appl Honey Phytochem 1:361–388

Rani GN, Budumuru R, Bandaru NR (2017) Antimicrobial activity of honey with special reference to methicillin resistant Staphylococcus aureus (MRSA) and methicillin sensitive Staphylococcus aureus (MSSA). J Clin Diagn Res 11(8):DC05

Ranneh Y, Akim AM, Hamid HA, Khazaai H, Fadel A, Zakaria ZA, Albujja M, Bakar MFA (2021) Honey and its nutritional and anti-inflammatory value. BMC Complement Med Ther 21(1):1–17. https://doi.org/10.1186/s12906-020-03170-5

Rao PV, Krishnan KT, Salleh N, Gan SH (2016) Biological and therapeutic effects of honey produced by honey bees and stingless bees: a comparative review. Rev Bras Farm 26:657–664

Rayhan MA, Yousuf SA, Rayhan J, Khengari EM, Nawrin K, Billah MM (2019) Black seed honey—a powerful ingredient of prophetic medicine; its neuropharmacological potential. J Apither 5(2):18–26

Razdar S, Panahi Y, Mohammadi R, Khedmat L, Khedmat H (2023) Evaluation of the efficacy and safety of an innovative flavonoid lotion in patients with haemorrhoid: a randomised clinical trial. BMJ Open Gastroenterol 10(1):e001158

Ren J, Cheng H, Xin W, Chen X, Hu K (2012) Induction of apoptosis by 7- piperazinethylchrysin in HCT-116 human colon cancer cells. Oncol Rep 28:1719–1726

Rickard MJ (2005) Anal abscesses and fistulas. ANZ J Surg 75(1–2):64–72

Rishton GM (2008) Natural products as a robust source of new drugs and drug leads: past successes and present day issues. Am J Cardiol 101(10):S43–S49

Safii SH, Tompkins GR, Duncan WJ (2017) Periodontal application of manuka honey: antimicrobial and demineralising effects in vitro. Int J Dent 2017:9874535. https://doi.org/10.1155/2017/9874535

Samarghandian S, Afshari JT, Davoodi S (2011) Chrysin reduces proliferation and induces apoptosis in the human prostate cancer cell line pc-3. Clinics 66(6):1073–1079

Samarghandian S, Farkhondeh T, Samini F (2017) Honey and health: a review of recent clinical research. Pharm Res 9(2):121

Sayed SF (2023) Herbal drugs as antibiotics. In: Dhara AK, Nayak AK, Chattopadhyay D (eds) Antibiotics-therapeutic spectrum and limitations. Academic Press, pp 479–532

Scepankova H, Saraiva JA, Estevinho LM (2017) Honey health benefits and uses in medicine. Bee Prod-Chem Biol Properties 83-96. https://doi.org/10.1007/978-3-319-59689-1_4

Scepankova H, Combarros-Fuertes P, Fresno JM, Tornadijo ME, Dias MS, Pinto CA et al (2021a) Role of honey in advanced wound care. Molecules 26(16):4784

Scepankova H, Pinto CA, Paula V, Estevinho LM, Saraiva JA (2021b) Conventional and emergent technologies for honey processing: a perspective on microbiological safety, bioactivity, and quality. Comp Rev Food Sci Food Saf 20(6):5393–5420

Schieber K, Niecke A, Geiser F, Erim Y, Bergelt C, Büttner-Teleaga A, Maatouk I, Stein B, Teufel M, Wickert M, Wuensch A, Weis J (2019) The course of cancer-related insomnia: don't expect it to disappear after cancer treatment. Sleep Med 58:107–113. https://doi.org/10.1016/j.sleep.2019.02.018

Schmidt B, Ribnicky DM, Poulev A, Logendra S, Cefalu WT, Raskin I (2008) A natural history of botanical therapeutics. Metabolism 57:S3–S9

Schneider LA, Korber A, Grabbe S et al (2007) Influence of pH on wound-healing: a new perspective for wound-therapy. Arch Dermatol Res 298(9):413–420

Schultz GS, Chin GA, Moldawer L, Diegelmann RF (2011) Principles of wound healing. In: Fitridge R, Thompson M (eds) Mechanisms of vascular disease: a reference book for vascular specialists. University of Adelaide Press, Adelaide (AU)

Šedík P, Pocol CB, Horská E, Fiore M (2019) Honey: food or medicine? A comparative study between Slovakia and Romania. Br Food J 121(6):1281–1297

Semprini A, Singer J, Braithwaite I, Shortt N, Thayabaran D, McConnell M et al (2019) Kanuka honey versus acyclovir for the topical treatment of herpes simplex labialis: a randomised controlled trial. BMJ Open 9(5):e026201. https://doi.org/10.1136/bmjopen-2018-026201

Shabani T, Hosseini SE (2017) Comparison of the hydro-alcoholic extract of ginseng root with metformin in rats with polycystic ovary syndrome. Armaghane Danesh J 21(11):1087–1099

Shahjehan RD, Bhutta BS (2023) Coronary artery disease. In: StatPearls. StatPearls Publishing, Treasure Island, FL

Shahrajabian MH, Sun W, Cheng Q (2021) Plant of the Millennium, Caper (*Capparis spinosa* L.), chemical composition and medicinal uses. Bullet Natl Res Centre 45:1–9

Shaito A, Thuan DTB, Phu HT, Nguyen THD, Hasan H, Halabi S et al (2020) Herbal medicine for cardiovascular diseases: efficacy, mechanisms, and safety. Front Pharmacol 11:422

Sharpe RM, Skakkebaek NE (1993) Are estrogens involved in falling sperm counts and disorders of the male reproductive tract? Lancet 341(8857):1392–1395

Shaw TJ, Martin P (2016) Wound repair: a showcase for cell plasticity and migration. Curr Opin Cell Biol 42:29–37

Sidik K, Mahmood A, Salmah I (2006) Acceleration of wound healing by aqueous extract of *Allium sativum* in combination with honey on cutaneous wound healing in rats. Int J Mol Med Adv Sci 2:231–235

Singh O, Khanam Z, Misra N, Srivastava MK (2011) Chamomile (*Matricaria chamomilla* L.): an overview. Pharmacogn Rev 5(9):82–95. https://doi.org/10.4103/0973-7847.79103

Singh A, Zhao K (2017) Treatment of insomnia with traditional Chinese herbal medicine. Int Rev Neurobiol 135:97–115

Sinha S, Sehgal A, Ray S, Sehgal R (2023) Benefits of Manuka Honey in the management of infectious diseases: recent advances and prospects. Mini Rev Med Chem 23(20):1928–1941

Skakkebaek NE, Rajpert-De Meyts E, Buck Louis GM, Toppari J, Andersson AM, Eisenberg ML, Jensen TK, Jørgensen N, Swan SH, Sapra KJ, Ziebe S (2016) Male reproductive disorders and fertility trends: influences of environment and genetic susceptibility. Physiol Rev 96(1):55–97

Smith ZL, Werntz RP, Eggener SE (2018) Testicular cancer: epidemiology, diagnosis, and management. Med Clin North Am 102(2):251–264

Sood SK, Bhardwaj R, Lakhanpal TN (2005) Ethnic Indian plants in cure of diabetes. Scientific Publishers

Sowa P, Grabek-Lejko D, Wesołowska M, Swacha S, Dżugan M (2017) Hydrogen peroxide-dependent antibacterial action of *Melilotus albus* honey. Lett Appl Microbiol 65(1):82–89

Spoială A, Ilie CI, Ficai D, Ficai A, Andronescu E (2022) Synergic effect of honey with other natural agents in developing efficient wound dressings. Antioxidants 12(1):34. https://doi.org/10.3390/antiox12010034

Stephen-Haynes J, Thompson G (2007) The different methods of wound debridement. Br J Community Nurs 12:S6–S16

Storgaard L, Bonde JP, Olsen J (2006) Male reproductive disorders in humans and prenatal indicators of estrogen exposure: a review of published epidemiological studies. Reprod Toxicol 21(1):4–15. https://doi.org/10.1016/j.reprotox.2005.05.006

Strothmann AL, Berne MEA, de Almeida Capella G, de Moura MQ, da Silva Terto WD, da Costa CM, Pinheiro NB (2022) Antiparasitic treatment using herbs and spices: a review of the literature of the phytotherapy. Braz J Vet Med 44

Subapriya R, Nagini S (2005) Medicinal properties of neem leaves: a review. Curr Med Chem Anticancer Agents 5(2):149–156

Subramanian AP, John AA, Vellayappan MV, Balaji A, Jaganathan SK, Mandal M, Supriyanto E (2016) Honey and its phytochemicals: plausible agents in combating colon cancer through its diversified actions. J Food Biochem 40(4):613–629

Sung H, Ferlay J, Siegel RL, Laversanne M, Soerjomataram I, Jemal A, Bray F (2021) Global cancer statistics 2020: GLOBOCAN estimates of incidence and mortality worldwide for 36 cancers in 185 countries. CA Cancer J Clin 71:209–249. https://doi.org/10.3322/caac.21660

Tachjian A, Maria V, Jahangir A (2010) Use of herbal products and potential interactions in patients with cardiovascular diseases. J Am Coll Cardiol 55(6):515–525

Takeda T, Sakata M, Isobe A et al (2008) Relationship between metabolic syndrome and uterine leiomyomas: a case-control study. Gynecol Obstet Invest 66(1):14–17

Takeo M, Lee W, Ito M (2015) Wound healing and skin regeneration. Cold Spring Harb Perspect Med 5:a023267

Takhtadzhian AL (1997) Diversity and classification of flowering plants. Columbia University Press

Talebi M, Talebi M, Farkhondeh T, Samarghandian S (2020) Molecular mechanism-based therapeutic properties of honey. Biomed Pharmacother 130:110590

Tandara A, Mustoe T (2004) Oxygen in wound healing-more than a nutrient. World J Surg 28:294–300

Thompson WG, Longstreth GF, Drossman DA, Heaton KW, Irvine EJ, Muller-Lissner SA (1999) Functional bowel disorders and functional abdominal pain. Gut 45(Suppl. 2):II43–II47. https://doi.org/10.1136/gut.45.2008.ii43

Thorup J, Cortes D, Petersen BL (2006) The incidence of bilateral cryptorchidism is increased and the fertility potential is reduced in sons born to mothers who have smoked during pregnancy. J Urol 176:734–737

Tierra M (1998) The way of herbs. Simon and Schuster

Tomás-Barberán FA, Ferreres F, Tomás-Lorente F, Ortiz A (1993) Flavonoids from Apis mellifera beeswax. Z Naturforsch C 48(1-2):68–72

Tomczyk M, Miłek M, Sidor E, Kapusta I, Litwińczuk W, Puchalski C, Dżugan M (2020) The effect of adding the leaves and fruits of Morus alba to rape honey on its antioxidant properties, polyphenolic profile, and amylase activity. Molecules 25:84. https://doi.org/10.3390/molecules25010084

Ullah A, Munir S, Badshah SL, Khan N, Ghani L, Poulson BG, Emwas A-H, Jaremko M (2020) Important flavonoids and their role as a therapeutic agent. Molecules 25(22):5243

Vaishnavi G, Rao M, Venkatesh P, Hepcykalarani D, Prema R (2020) A review on anesthetic herbs. Res J Pharmacogn Phytochem 12(1):52–56

Van der Heijden JF, Hassink RJ (2013) The phospholamban p. Arg14del founder mutation in Dutch patients with arrhythmogenic cardiomyopathy. Netherlands Heart Journal 21:284–285

Van Oudenhove L, Vandenberghe J, Demyttenaere K, Tack J (2010) Psychosocial factors, psychiatric illness and functional gastrointestinal disorders: a historical perspective. Digestion 82:201–210

Variya BC, Bakrania AK, Patel SS (2016) Emblica officinalis (Amla): a review for its phytochemistry, ethnomedicinal uses and medicinal potentials with respect to molecular mechanisms. Pharmacol Res 111:180–200

Velayutham P, Babu PV, Liu D (2008) Green tea catechins and cardiovascular health: an update. Curr Med Chem 15(18):1840–1850

Verma RK, Sharma N (2022) Phytochemical and pharmacological activities of: a review *Alpinia galangal*. Asian J Pharm Pharmacol 8(3):74–85

Vlcekova P, Krutakova B, Takac P, Kozanek M, Salus J, Majtan J (2012) Alternative treatment of gluteo femoral fistulas using honey: a case report. Int Wound J 9(1):100–103. https://doi.org/10.1111/j.1742-481X.2011.00844.x

Vogt D, Borowski SC, Godier-McBard LR, Fossey MJ, Copeland LA, Perkins DF, Finley EP (2022) Changes in the health and broader well-being of U.S. veterans in the first three years after leaving military service: overall trends and group differences. Soc Sci Med 294:114702. https://doi.org/10.1016/j.socscimed.2022.114702

Walia K, Boolchandani R, Dhand S, Antony B (2015) Improving glycemic & lipidemic profile with amla powder (*Emblica officinalis*) supplementation in adults with type 2 diabetes mellitus. Int J Basic Appl Med Sci 5:251–258

Walsh B (2004) Asia's war with heart disease. Time Asia 10:501040510–501632135

Waykar B, Alqadhi YA (2016) Biological properties and uses of honey: a concise scientific review. Indian J Pharm Biol Res 4(3):58–68

Weber MA, Schiffrin EL, White WB, Mann S, Lindholm LH, Kenerson JG et al (2014) Clinical practice guidelines for the management of hypertension in the community: a statement by the American Society of Hypertension and the International Society of Hypertension. J Clin Hypertension 16(1):14

Wee JJ, Chung AS (2012) Biological activities of ginseng and its application to human health. In: Herbal medicine: biomolecular and clinical aspect, 2nd edn. CRC Press

Weerakkody NS, Caffin N, Lambert LK, Turner MS, Dykes GA (2011) Synergistic antimicrobial activity of galangal (*Alpinia galanga*), rosemary (*Rosmarinus officinalis*) and lemon iron bark (*Eucalyptus staigerana*) extracts. J Sci Food Agric 91(3):461–468

Weinberg RA (1989) Oncogenes, anti-oncogenes, and the molecular basis of multistep carcinogenesis. Cancer Res 49:3713–3721

Wilkinson HN, Hardman MJ (2020) Wound healing: cellular mechanisms and pathological outcomes. Open Biol 10:200223. https://doi.org/10.1098/rsob.200223

Wink M (2015) Modes of action of herbal medicines and plant secondary metabolites. Medicines 2(3):251–286

Xie W, Fu X, Tang F, Mo Y, Cheng J, Wang H, Chen X (2019) Dose-dependent modulation effects of bioactive glass particles on macrophages and diabetic wound healing. J Mater Chem B 7:940–952

Yaacob N, Nengsih A, Norazmi M (2013) Tualang honey promotes apoptotic cell death induced by tamoxifen in breast cancer cell lines. Evid Based Complement Alternat Med 2013:989841. https://doi.org/10.1155/2013/989841

Yaghoobi N, Al-Waili N, Ghayour-Mobarhan M, Parizadeh S, Abasalti Z, Yaghoobi Z, Yaghoobi F, Esmaeili H, Kazemi-Bajestani S, Aghasizadeh R (2008) Natural honey and cardiovascular risk factors; effects on blood glucose, cholesterol, triacylglycerole, CRP, and body weight compared with sucrose. Sci World J 8:463–469

Yamin M, Ayu DF, Hamzah F (2017) Lama pengeringanterhadapaktivitasantioksidan dan mututeh herbal daunketapengcina (Cassia alataL). Jom FAPERTA 4(2):1–15

Yang Y, Karakhanova S, Werner J, Bazhin A (2013) Reactive oxygen species in cancer biology and anticancer therapy. Curr Med Chem 20:3677e92

Yi LV, Kwok-Fai SO, Nai-Kei WONG, Jia XIAO (2019) Anti-cancer activities of S-allylmercaptocysteine from aged garlic. Chin J Nat Med 17(1):43–49

Yoo DY, Kim W, Nam SM, Yoo M, Lee S, Yoon YS et al (2014a) Neuroprotective effects of Z-ajoene, an organosulfur compound derived from oil-macerated garlic, in the gerbil hippocampal CA1 region after transient forebrain ischemia. Food Chem Toxicol 72:1–7

Yoo M, Lee S, Kim S, Hwang JB, Choe J, Shin D (2014b) Composition of organosulfur compounds from cool-and warm-type garlic (*Allium sativum* L.) in Korea. Food Sci Biotechnol 23:337–344

Zaid SSM, Sulaiman SA, Sirajudeen KNM, Othman NH (2010) The effects of tualang honey on female reproductive organs, tibia bone and hormonal profile in ovariectomised rats-animal model for menopause. BMC Complement Alternat Med 10:82. https://doi.org/10.1186/1472-6882-10-82

Zaidi A, Green L (2019) Physiology of haemostasis. Anaesth Intens Care Med 20:152–158

Zamanian M, Azizi-Soleiman F (2020) Honey and glycemic control: a systematic review. Pharma Nutr 11:100180

Zdravkovic S et al (2002) Heritability of death from coronary heart disease: a 36-year follow-up of 20 966 Swedish twins. J Intern Med 252:247–254

Zhang CM, Fan PH, Li M, Lou HX (2014) Two new sesquiterpenoids from the rhizomes of *Curcuma xanthorrhiza*. Helv Chim Acta 97(9):1295–1300

Zhu S, Viejo-Borbolla A (2021) Pathogenesis and virulence of herpes simplex virus. Virulence 12(1):2670–2702. https://doi.org/10.1080/21505594.2021.1982373

Zohrevand Asl Z, Hayati Roudbari N, Mohamadi Gorji S, Parivar K (2017) The effect of red ginseng water extract on oogenesis and uterus tissue of adult NMR mice. J Anim Biol 9(3):63–75

Zuo W, Yan F, Zhang B, Li J, Mei D (2017) Advances in the studies of *Ginkgo biloba* leaves extract on aging-related diseases. Aging Dis 8(6):812

Printed in the United States
by Baker & Taylor Publisher Services